Kelvin and Ireland

Journal of Physics: Conference Series Volume 158

Belfast, Northern Ireland, UK
7 December 2007

ISBN: 978-1-61738-385-4

Printed from e-media with permission by:

Curran Associates, Inc.
57 Morehouse Lane
Red Hook, NY 12571

CURRAN ASSOCIATES INC. proceedings.com

Some format issues inherent in the e-media version may also appear in this print version.

Printed by Curran Associates, Inc. (2010)

Institute of Physics
Dirac House, Temple Back
Bristol BS1 6BE UK

Phone: 44 1 17 929 7481
Fax: 44 1 17 920 0979

techtracking@iop.org

Additional copies of this publication are available from:

Curran Associates, Inc.
57 Morehouse Lane
Red Hook, NY 12571 USA
Phone: 845-758-0400
Fax: 845-758-2634
Email: curran@proceedings.com
Web: www.proceedings.com

TABLE OF CONTENTS

Sir Joseph Larmor unveiling the Kelvin memorial in the Botanic Gardens, Belfast on a rainy day in 1913

Preface

William Thomson, later Lord Kelvin, was born in Belfast in 1824, and his family had lived near Ballynahinch in the north of Ireland, quite close to Belfast, from the seventeenth century. At the time of Kelvin's birth, James Thomson, his father, was Professor of Mathematics at the Belfast Royal Academical Institution (Inst). However, following the death of his wife in 1830, James took up a new position as Professor at the University of Glasgow, and he and his children moved there in 1832.

Apart from three years studying at Cambridge, and a very brief period immediately afterwards travelling and teaching in Cambridge, Kelvin was to spend the rest of his life in Glasgow, where he occupied the Chair of Natural Philosophy (or Physics) for 53 years. The natural assumption might be that his birth in Ireland was irrelevant to Kelvin's life and work, and that the fine monument erected in his honour in Belfast's Botanic Gardens, which is pictured on the front cover of this volume, was more a demonstration of civic pride than a recognition of an aspect of Kelvin's life which was important to him.

The purpose of the meeting was to demonstrate that this was not the case, that, great Glaswegian as he undoubtedly became, Kelvin always delighted in the title of Irishman. The influence of his father, very much an Ulsterman, was immense, and Kelvin and his siblings were to follow his non-sectarian and reforming approach. Also important for Kelvin was his Christian upbringing, which began in Belfast, and his beliefs were to play a role of importance in his life and indeed in much of his most important work, in particular that on thermodynamics.

Two of his siblings returned to Belfast and spent much of their lives there, and Kelvin was a frequent visitor. While the most celebrated example of his engineering work was his massive contribution to the Atlantic telegraph cable, and his maritime and other inventions were used throughout the world, he was especially pleased to contribute to two important technical accomplishments in the north of Ireland, the occulting Holywood lighthouse and the Giant's Causeway tramway. Kelvin's interest in Ireland came again to the fore in the Home Rule debate at the end of the nineteenth century; Kelvin was vehemently opposed to Home Rule and became a passionate member of the Liberal Unionist party, a crucial factor in his becoming a peer.

Accounts of these aspects of Kelvin's involvement with Ireland are given in our first four papers. The fifth paper was particularly appropriate for the meeting. It describes how an important contribution of Kelvin on optimal packing has been improved in Ireland by the speaker, Denis Weaire, himself, and how the result has been seen by the whole world in Beijing.

Finally there is a discussion of Kelvin's place in the story of Irish science. There have been many highly accomplished Irish scientists, and the meeting showed clearly that Kelvin certainly deserves to be included among the greatest.

Acknowledgments

The meeting was the second regional event of the Institute of Physics (IOP) in Ireland, the first being the meeting on *John Desmond Bernal: Science and Society* held in Limerick in June 2006 and published in the same series as this volume. The meeting also benefited from the support of many bodies and institutions.

We would like to thank the committee of the IOP in Ireland for providing finance and also giving their full support. We would particularly mention Vincent Casey, the Chair, who also chaired the afternoon session, Emma Sokell, the Secretary, and Denise Gabuzda and Kevin McGuigan, past and present Treasurers. We would also thank Sheila Gilheany and Alison Hackett for general help and encouragement, including assisting the organisation on the day of the meeting.

We also acknowledge financial support from the IOP in London, and in particular thank Susan Lippmann and John Brindley. We also thank Peter Ford, present Chair of the History of Physics Group at the IOP for general encouragement and helping us to obtain this finance.

The meeting was also supported by Queen's University Belfast. We would thank Tom Millar, Dean of the Faculty of Engineering and Physical Science for arranging this support. We were supremely fortunate to be allowed to use the magnificent Great Hall, a highly suitable room for the meeting since among the portraits on the walls were many of those referred to in the talks, and even one of the speakers! Catering and audiovisual services were also very helpful. From the School of Mathematics and Physics at Queen's, we would acknowledge the considerable help of Angela Anderson and Margaret Hewitt.

We would also thank Kevin Mulhern, who arranged for publicity, Lisa Mitchell who provided the publicity in a splendid way, and also Eugene McCusker and Claire Wilson. Juliet Chantler of Northern Visions prepared an enthralling TV programme around the event and we are very grateful to her also. Joe Millar of the Ulster Scots Agency also publicised the meeting and we thank him as well.

David Livingstone gave general support and chaired the morning session, for which we are grateful, and of course we would like to thank all the speakers who were responsible for the meeting being interesting and enjoyable. We would mention Andrew Holmes who gave a very erudite paper at the meeting. Unfortunately, since his material is being published elsewhere, we have not been able to include his contribution here, but he has given helpful advice on the preparation of this volume.

Finally we would like to thank Graham Douglas, Publisher for the Journal of Physics Conference Series for his courtesy and helpfulness in the preparation of this volume.

Figure Credits

The Ulster Connection (Raymond Flood)

Figure 2. Taken from S.P. Thompson, The Life of William Thomson Baron Kelvin of Largs, 2 volumes, Macmillan & Co., London, 1910, opposite p164.

Figure 3. Taken from James Thomson, An integrating machine having a new kinematic principle, Transactions of the Royal Society, **24**, 262-5, 1876.

Figure 4. Taken from Sir William Thomson, Popular Lectures and Addresses, vol. III, Macmillan & Co., London, 1891, p184.

Figure 6. Taken from Nature, **6**, May 16th ,1872, p46.

Figure 7. Taken from C.G. Knott, Life and Scientific Work of P.G. Tait, Cambridge University Press, 1910, opposite p325.

Figure 8. Taken from A.G. King, Kelvin the Man, Hodder & Stoughton, London, 1925, opposite p104.

From a land of saints and scholars? Reflections on William Thomson's religious background (Mark McCartney)

Figure 1 Taken from A. Russell, Lord Kelvin, Blackie & Son Ltd, Glasgow, 1938.

Figure 2. Taken from R Glazebrook , Kelvin Centenary Oration and Addresses Commemorative, Percy Lund Humphries & Co Ltd, London, 1924, p76.

Figure 3. Image courtesy of Capt. J.L. Davidson, Lamlash, Isle of Arran.

Figure 4. Image courtesy of, and copyright owned by, S.K. Houston.

Kelvin and industry in Ireland (Bernard Crossland and Andrew Whitaker)

Figure 1. http://zapatopi.net/kelvin/papers/lighthouses_of_the_future.html; article from *Good Words* 1873, pages 217-224.

Figure 2. Picture owned by and reproduced by permission of the Ulster Museum; Welch collection,

Figure 3. Picture owned by and reproduced by permission of Giant's Causeway and Bushmills Railway.

Lord Kelvin and Liberal Unionism (Iain Hutchison)

Figures 1-12, 17, 22-25. Wikimedia Commons; GNU Free Documentation License. In the public domain because copyright has expired.

Figures 11, 13-16, 20-21. *The Bailie* (published in Glasgow; ceased publication in April 1926; publisher long defunct).

Kelvin's ideal foam structure (Denis Weaire)

Figures 1 and 2. Courtesy of the author.

Is Kelvin representative of an "Irish tradition"? (Andrew Whitaker)

Figures 1-9, 13-17, 20-26, 28-30, 34, 39-44, 47-48, 50, 54-55, 59-62, 64-66, 68, 72-74. Wikimedia Commons; GNU Free Documentation License. In the public domain because copyright has expired

Figures 10, 12, 19 and 53. Owned by and used by permission of Queen's University Belfast.

Figure 11. http://www.victorianweb.org/sculpture/misc/cooke1.html

Figure 18. Picture owned by and reproduced by permission of Trinity College Dublin.

Figure 27 http://www.speedace.info/index.htm ; reproduced by permission of Electrick Publications and NJK.

Figure 31. Picture owned by and reproduced by permission of the National Portrait Gallery, London.

Figures 32 and 33. Pictures owned by and reproduced by permission of the USCC Special Collections; Lick Observatory Records.

Figures 35 and 51. Picture owned by and reproduced by permission of Engineers Ireland.

Figure 36. Picture owned by and reproduced by permission of the London Mathematical Society.

Figures 37 and 57. Pictures owned by and reproduced by permission of the Royal Dublin Society.

Figure 38. Picture owned by and reproduced by permission of the Royal Society.

Figure 45. Picture owned by and reproduced by permission of The Irish Picture Collection/Birr Castle Collection.

Figures 46 and 49. Pictures owned by and reproduced by permission of the University of London Observatory: University of London.

Figures 52 and 67. From www.universitystory.gla.ac.uk and reproduced by permission of the Glasgow University Archive Services.

Figure 56. Picture owned by and reproduced by permission of Maynooth College.

Figure 58. From Obituary in Physical Review **11**, 188 (1914).

Figure 63. From: The Scientific Life and Influence of Clifford Ambrose Truesdell III, Archive for Rational and Mechanical Analysis 161, 1-26 (2002); image reproduced by permission of Professor Ball.

Figures 69 and 70. Pictures owned by and reproduced by permission of the Clarendon Laboratory

Figure 71. Picture owned by and reproduced by permission of the Museum of History of Science, Oxford.

Figure 75. Picture owned by and reproduced by permission of Nottingham City Council Archive.

The editors and authors have attempted to trace the copyright holders of all the figures produced in this publication and apologise to them if permission to publish in this form has not been obtained.

The Ulster connection

Raymond Flood
Department for Continuing Education, University of Oxford, Oxford OX1 2JA England

E-mail: raymond.flood@continuing-education.oxford.ac.uk

Abstract. This article will include an account of Kelvin's father, James Thomson, whose first job was as a farm labourer near Belfast, Northern Ireland where Kelvin was born. James went on to be Head of Mathematics at Belfast Academical Institution, and then Professor of Mathematics at Glasgow University. He played an enormously influential role in the development of Kelvin's religious, scientific and political views, as well as being central to Kelvin also obtaining a Chair at Glasgow. The talk will also include an account of Kelvin's personal interaction with his elder brother, also called James, who was an engineer and returned to work in Belfast in 1851, being Professor of Engineering at Queen's University, Belfast, from 1854 until he moved to join Kelvin in Glasgow in 1873. Kelvin's sisters also returned to live in Belfast, and he was himself a frequent visitor, always concerned with Belfast and Irish affairs. December 2007 marks the centenary of Kelvin's death and this is an appropriate point to review the Irish dimension of his life so frequently overlooked.

1. Introduction

On the 23rd of December 1907 William Thomson, Baron Kelvin of Largs was buried in Westminster Abbey beside the nation's most venerated scientist, Sir Isaac Newton. In his lifetime Thomson was seen as the most important natural philosopher of the Victorian age, but the waxing and waning of the twentieth century and the supplanting of classical physics have eroded his reputation, so that for many scientists he is remembered as little more than a unit of temperature.

This article draws on the book *Kelvin: Life, Labours and Legacy* edited by myself, Mark McCartney, and Andrew Whitaker [1]. It is a book of essays that seeks to re-evaluate and rehabilitate Lord Kelvin, setting both the man and his work in historical and scientific context. It is not of course the first book to do so but the centenary of his death seems an appropriate to revisit his life, labours and legacy. This talk draws particularly on the chapters by Mark McCartney, Peter Bowler, Alex Craik, Andrew Whitaker and myself.

It was a family joke that William Thomson had two birthdays, and once while signing his name in a birthday book, as William Thomson on the 25th June, and Kelvin on the 26th he quipped "It is convenient to have two birthdays when you have two names!" It is not quite clear how the confusion arose, and though in childhood his birthday was celebrated on the 25th we have it on the authority of his father that William Thomson was born on the 26th June 1824 at 5 a.m

Due to the influence and attitude of their father, Dr James Thomson, William and his brothers and sisters enjoyed a happy childhood, in a supportive and learning environment. His father was born in 1786 in a farmhouse outside Ballynahinch, County Down and he was the youngest – by 10 years - of five

Journal of Physics: Conference Series **158** (2009) 012001 doi:10.1088/1742-6596/158/1/012001

children. The family was descended from seventeenth-century Scottish settlers. Retaining their Presbyterian religion, this large contingent of immigrants, sometimes known as "Ulster Scots", formed a distinct strand of Irish society. Few were rich, and many later emigrated to the United States of America. His two older sisters taught him to read and he taught himself arithmetic from a copy of John Bonnycastle's *The Scholar's Guide to Arithmetic*. He went on to study at Glasgow University, and by 1815 he was Professor of Mathematics at the newly opened Belfast Academical Institution (today a school, but in the early days of its existence a school and college combined, with the college functioning like a small Scottish university).

He was a successful writer of textbooks - on topics such as arithmetic, calculus, trigonometry and geography (rising at 4am in the morning to work on them). Note (figure 1) that the treatise on Arithmetic went to at least 72 editions. In fact this was the last edition and was revised by his sons, James and William. The first edition was in 1819 so it had 72 editions over a sixty year period.

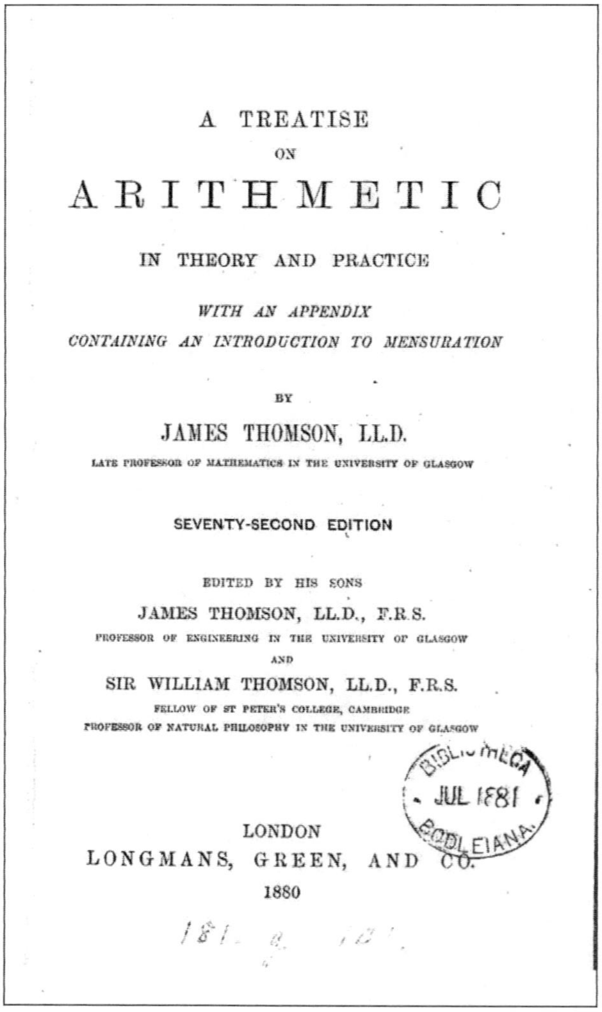
A TREATISE

ON

ARITHMETIC

IN THEORY AND PRACTICE

WITH AN APPENDIX
CONTAINING AN INTRODUCTION TO MENSURATION

BY

JAMES THOMSON, LL.D.

LATE PROFESSOR OF MATHEMATICS IN THE UNIVERSITY OF GLASGOW

SEVENTY-SECOND EDITION

EDITED BY HIS SONS

JAMES THOMSON, LL.D., F.R.S.

PROFESSOR OF ENGINEERING IN THE UNIVERSITY OF GLASGOW

AND

SIR WILLIAM THOMSON, LL.D., F.R.S.

FELLOW OF ST PETER'S COLLEGE, CAMBRIDGE
PROFESSOR OF NATURAL PHILOSOPHY IN THE UNIVERSITY OF GLASGOW

LONDON
LONGMANS, GREEN, AND CO.
1880

Figure 1. The 72nd, and final, edition of Dr James Thomson's Treatise on Arithmetic, edited by his sons James and William.

James Thomson described his aim in the preface as "to present a full and regular course of whatever is useful in arithmetic."

The book uses three different sizes of type to distinguish between important definitions and rules compared to principal illustrations, with the less important illustrations in the smallest type. Using different sizes of type was a technique that his son William also used with P.G. Tait in their influential "Treatise on Natural Philosophy".
James Thomson senior also chose practical examples to illustrate the subject. As he wrote

> "Of the Examples and Exercises, some are proposed in purely abstract terms, being intended to merely to afford practice to the learner in the rules and modes of calculation. To these are subjoined, in those parts of the work in which it could be conveniently done, other questions, which will not only afford the pupil farther exercise on the rules which precede them, but will also furnish him with many important facts in Commerce, Geography, Astronomy, Chronology, Chemistry, and other departments of knowledge…. The information contained in these questions has been all derived from authentic sources…" [2]

This emphasis on practical matters and applications was characteristic of James Thomson's approach and influenced his sons, James and William.
James's hard work and ambition paid off and opposite the Academical Institution he built two houses, one to raise his family in, and one to rent out. Elizabeth (born 1818) was the first of James and his wife Margaret's 7 children – William was the middle child. In her reminiscences Elizabeth describes a happy and affectionate childhood in Belfast
James's brother Robert, who still farmed the land outside Ballynahinch wrote to his sister of their Belfast nephews and nieces that they were

> "wonderfully apt in learning; but I don't think it strange, as both father and mother are drilling them." [3]

In 1830 the family suffered the blow of their mother's death. Two years after this, the family moved to Glasgow, where James Thomson had been appointed Professor of Mathematics at the university.
Joining Professor Thomson in his Junior Mathematics class in the 1832-33 academic year were his two elder boys James and William. It can be confusing as there were two James, James the father and James the brother.
They attended the class, but were not formally enrolled, and did not take examinations. However, in the 1834-35 session, with James now aged 12 and William 10, they did enroll. This is remarkable, but not quite as remarkable as it might first appear, as the usual minimum age for matriculation at Glasgow was 14. They took prizes in a range of classes throughout their time there. However most consistent was their performance in the mathematics and natural philosophy classes, where William regularly took first prize and James second.
Dr James Thomson's hard work paid off financially, and by 1839 he was wealthy enough to take the family to the continent for two months during the summer. As part of their journey they traveled from Glasgow to Liverpool by boat and from Liverpool to London by train – "Going at a tremendous rate- no less than thirty-six miles an hour!" - reports his sister, Elizabeth [4]. The family spent two weeks in Paris together and then James left the boys in Paris with the family servant to have daily lessons in French. While there William went to the Bibliothèque Royale to read Laplace's *Mécanique Céleste* in preparation

Journal of Physics: Conference Series **158** (2009) 012001 doi:10.1088/1742-6596/158/1/012001

for an 85 page and mathematically sophisticated essay *On the Figure of the Earth* which won him a University medal in his next academic year.

The next year it was Germany. In spite of his father's instruction to concentrate on learning German, Thomson, by now just 16, had packed a little French reading: Fourier's *Théorie Analytique de la Chaleur*. William Thomson continues the story:

> "Going that summer to Germany with my father and my brothers and sisters, I took Fourier with me. My father took us to Germany and insisted that all work should be left behind, so that the whole of our time could be given to learning German. We went to Frankfort, where my father took a house for two months… Now just two days before leaving Glasgow I had got Kelland's book (Theory of Heat, 1837), and was shocked to be told that Fourier was mostly wrong. So I put Fourier in my box, and used in Frankfort to go down to the cellar surreptitiously every day to read a bit of Fourier. When my father discovered it he was not very severe upon me." [5]

Not a typical concern of a father to stop a child doing homework! As a result of his studies William claimed that Philip Kelland, the Professor of Mathematics at Edinburgh was wrong in his criticisms of Fourier. Although initially incredulous of his son's claims, a closer examination showed that they were accurate, and Thomson wrote the matter up in what was to be his first published paper appearing in May 1841 in the Cambridge Mathematical Journal, with the author simply designated as "P.Q.R."

2. Cambridge

Thomson's strength in mathematics was obvious, and the clear place for him to study after finishing Glasgow was Cambridge. October 1841 saw William Thomson, aged 17, enter St Peter's College (Peterhouse) Cambridge as a pensioner i.e. a student who paid his own way. During his time there the Cambridge Mathematical Journal published a further 10 papers from William Thomson - all using pseudonyms, with all but one under that of P.Q.R. Indeed, Thomson was being tipped as Senior Wrangler virtually as soon as he arrived in Cambridge.

At Cambridge there was regular and frequent correspondence with his family. His father, who was footing the considerable bills wrote to him of the wise use of time and money and care in the friendships he cultivated: "Recollect my invaluable maxim never to quarrel with a man (but to waive the subject) about religion." [6] And "Use all economy consistent with respectability. Be most circumspect about your conduct and about what acquaintance you form. You are young: take care you be not led to what is wrong." [7]

Dr James Thomson's worries were not unreasonable. Apparently many a young man came to Cambridge and filled his time with nothing but boating and wine parties. Thomson began his career at Cambridge with whole-hearted promises to his father: He had given no wine parties, and he had not joined the rowing club.

Initially he claimed that the college was divided between rowing men and reading men and that "rowing for the races is too hard work for getting on well with reading" [8], but by February 1842 he had bought a secondhand boat - a bargain at £7, and, he claimed, much cheaper than renting a boat. His father was unimpressed, but sent money to cover the bill. In spring of the next year his diary records late nights with a wide circle of friends, bathing, skating, walking, reading and serious rowing. He was also one of the founding members of the Cambridge University Musical Society.

Letters to his father are affectionate and confident in their relationship: "My Dear Father, I have again to write to you on the same pleasant business that I had to write to you so lately, which is to say that my money is again all gone." [9]

To soften his father up when asking for more money he would sometimes include a mathematical problem for use in the exams at Glasgow, or mention the receipt of another scholarship or prize (of which he received over a dozen during his student days at St Peter's). Over the three years from autumn 1841 to autumn 1844 Thomson's education had cost over £770. On sending his son a letter containing a summary of this cost Dr Thomson writes, with exasperation, "How is this to be accounted for? Have you lost money or been defrauded of it?... you must exercise the strictest economy… not spending a penny unnecessarily" [10]. To put this in context an upper middle class income at this period would have been around £900 – 1000 per annum.

William, however was working hard under the sharp eye of his private tutor William Hopkins. Hopkins' reputation as 'the senior wrangler maker' was well earned. By 1849 he had coached some 200 wranglers – a fifth of whom had landed one of the top three places.

Those in the top class were called Wranglers in an echo of the old system of disputation. The candidates were listed in order of marks with the top candidate being 'Senior Wrangler' the next 'Second Wrangler' and so on.

He was an excellent teacher, and a hard task master. The Mathematical Tripos was the most grueling academic race course in the country, and Hopkins trained and honed his candidates well. Thomson's examinations commenced on New Year's Day 1845, and finished on the 7th of January. There were 12 papers, with morning papers being two and a half hours long, and afternoon papers three hours long. There was a high proportion of 'bookwork' based questions and the papers required fluent recall, a good knowledge of mathematical tricks and shortcuts, and fast penmanship.

To universal surprise when the results were read out Thomson came second, with one Stephen Parkinson, a student at St. John's coming first. One explanation for Parkinson's victory is that, quite simply, he wrote much faster than Thomson. Indeed one of the examiners commented that if he had not seen Parkinson's speed in the examination hall he would have struggled to believe that the sheer volume of material could have been written by him in the time limit. However, when it came to the Smith's Prize examinations later in January Thomson easily took first place. These papers were more heavily weighted towards problem solving, and Thomson convincingly outperformed Parkinson

William Thomson's failure to be senior wrangler has generated at least two, possibly apocryphal, stories. The first, which has at least the provenance of being told by Joseph Larmor and being recorded in the official biography by Silvanus P. Thompson is that in the Smith's Prize papers the two top candidates presented solutions to a question which were so similar that the matter was investigated further. When questioned, the senior wrangler, Parkinson, said that the solution he had given came from a paper he had read in the Cambridge Mathematical Journal. The paper, which had been published some years earlier, had been authored by someone who signed himself simply as "P.Q.R.". – which, of course, was Thomson's pseudonym.

The second anecdote is that on the day the results of the Tripos were to be announced at the Senate House William dispatched his college servant with the words "Oh, just run down to the Senate House, will you, and see who is Second Wrangler." The servant returned and announced "You sir!" This seems unfair as Thomson does not seem to have been conceited.

Thomson's undergraduate success brought an election to a Fellowship of Peterhouse in June 1845 – he had just turned 21.

3. Glasgow

While Thomson was still a Cambridge undergraduate, events were unfolding at Glasgow which would determine his future career. William Meikleham had been the Professor of Natural Philosophy at Glasgow University since 1803, and during the academic session of 1838-39 his health became poor and his classes were subsequently covered by colleagues. By 1841 it was realized that Meikleham was unlikely ever to return to his classes and Dr James Thomson's thoughts turned to his colleague's eventual replacement. He wanted academic excellence, of the Cambridge calibre – a member of the intellectual elite, but not an elitist. The new member of staff needed to be in sympathy with the broad and non-hierarchical Scottish university education system and above all he had to be a good teacher.

There was a range of possible candidates, but by perhaps as early as Christmas 1842 James Thomson had realized that his son William, then only 18, could be in the running for the job. Certainly by March 1843 he was encouraging his son in a letter to cultivate particular friendships which could prove useful "in case of a certain event coming round". [11]

The next month he was gently testing his ideas out and one colleague identified a potential weakness, William's lack of experience in experimental work and William as a result attended lecture courses on experimental natural philosophy twice (once in 1843 & again in 1844), and in 1844 he attended lectures on practical astronomy and astronomical instruments. Also during the summer of 1843 he spent a month working in the chemistry labs at Glasgow.

Part of the maneuvering also included a four and a half month trip to Paris after graduation in 1845. In Paris he attended lectures on chemistry and physics at the Sorbonne. Through introductions provided by Scottish academics such as JD Forbes of Edinburgh and Sir David Brewster of St Andrews he met eminent men like Cauchy and Biot, and through Biot he was introduced to Victor Regnault, who was the professor of Natural Philosophy at the Collège de France. Regnault was happy for Thomson to assist in his laboratory. Thomson made use of the library and helped with experiments, observing what he described much later as Regnault's "faultless technique, a love of precision in all things, and the highest virtue of the experimenter - patience." [12]

Back in Glasgow Dr James Thomson was pleased to hear about his son's progress and networking. Not everyone was necessarily looking forward to the prospect of William Thomson joining his father as a professor. James sought change at Glasgow, and that change wasn't always welcomed by his colleagues. Appointing William might mean one more professorial vote in support of his agenda. In a remark to one of his children Dr James Thomson commented that at least one of his colleagues would rather see Satan appointed to the chair of Natural Philosophy than have William take the post.

Professor William Meikleham eventually died on 6th May 1846, and from that moment James Thomson's covert maneuverings turned to overt action. He wrote immediately to his son "The enclosed notice [of Meikleham's death] must put you into active and energetic motion without delay".[13]

Less than three weeks later on the 26th May, William Thomson wrote letters to each of the electors announcing his wish to be a candidate for the post, and then a long list of testimonials were collected to swamp and impress the Glasgow professors. The list reads like a selection from a Who's Who of 19th century science including George Boole, Arthur Cayley, James D. Forbes, Sir William Rowan Hamilton, Joseph Liouville, Agustus De Morgan, Victor Regnault, George Gabriel Stokes, James Joseph Sylvester and William Whewell.

When it came to the election itself none of Dr. James Thomson's fears were realized – possible candidates whom James had fretted over, and who would have been genuine competitors to his son simply did not apply. On the 11th September 1846, William Thomson, then aged 22, was unanimously elected to the Chair of Natural Philosophy at Glasgow. He held the post until 1899, a period of 53 years, and could not be persuaded to leave even by the Cavendish chair at Cambridge, which was offered to him three times, on its creation in 1870, again on Maxwell's death in 1879 and once more when Lord Rayleigh vacated it in 1884.

4. Collaboration between James and William

During the winter of 1848-49, Glasgow experienced an outbreak of cholera in which over 3500 people died. One of its victims was James Thomson senior, who died on January 12th, 1849 at the age of sixty-

Journal of Physics: Conference Series **158** (2009) 012001 doi:10.1088/1742-6596/158/1/012001

two. He and William, his son, had been Glasgow colleagues for only two years. He did not live to enjoy William's major successes. Also he did not live to see his eldest son, James, occupy the chairs of engineering at Queen's College Belfast (1857-73) and then Glasgow University (1873-89).

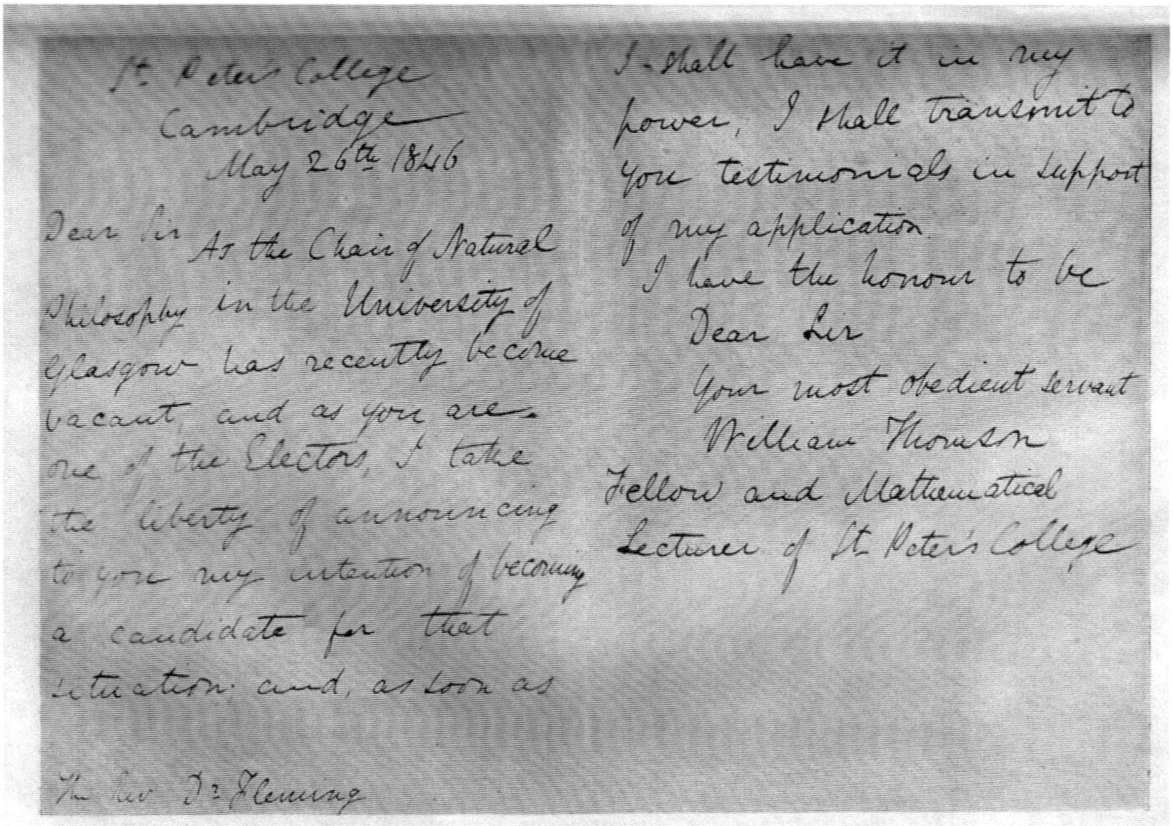

St Peter's College
Cambridge
May 26th 1846

Dear Sir As the Chair of Natural Philosophy in the University of Glasgow has recently become vacant, and as you are one of the Electors, I take the liberty of announcing to you my intention of becoming a candidate for that situation; and, as soon as I shall have it in my power, I shall transmit to you testimonials in support of my application.

I have the honour to be
Dear Sir
Your most obedient servant
William Thomson
Fellow and Mathematical Lecturer of St Peter's College

The Rev Dr Fleming

LETTER ANNOUNCING CANDIDATURE FOR THE CHAIR OF NATURAL PHILOSOPHY.

Figure 2. William Thomson announces his 'intention of becoming a candidate' for the Chair of Natural Philosophy at Glasgow.

As Peter Bowler states

> "What I want to suggest is that the two brothers, William and James, stand at opposite ends of a spectrum of interests by which science interacted with technology in the nineteenth century. William was a physicist with a strong interest in engineering, and James was an engineer with a strong interest in physics." [14]

William's elder brother James was born, two years before William in 1822 and worked under engineers in various parts of the country, ending up at the Millwall shipbuilding works of William Fairburn, one of the leading figures in the construction of the new ocean-going iron steamships. His health then broke down and he moved back to Glasgow where he collaborated with William in his work on thermodynamics

Journal of Physics: Conference Series **158** (2009) 012001 doi:10.1088/1742-6596/158/1/012001

and began his career as an inventive engineer. In 1851 he moved to Belfast where he opened an office as a civil engineer and served as engineer to the Water Commissioners. He married Elizabeth Hancock in 1853. The following year he became acting Professor of Engineering at Queen's, being appointed to the chair of engineering three years later. Thomson replaced James Godwin, the first Professor of Engineering, who had worked mainly on railways. He lived for some time at No. 17 University Square where he redesigned the sewers. He remained at Queen's until moving to take up the chair of engineering at Glasgow in 1873. He died in May 1892, followed within a week by his wife and younger daughter, all succumbing to a "severe cold", probably pneumonia.

James collaborated with William, for example in 1847 he predicted that the freezing point of water would be decreased if the pressure was increased which was shown experimentally by William in the following year. Another collaboration was to do with mechanical integration.

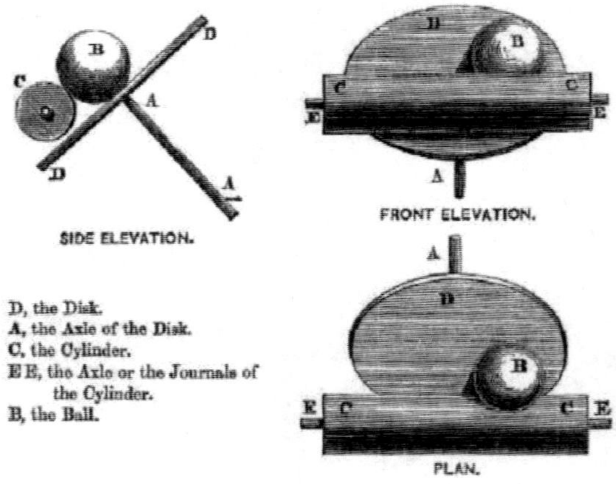

Figure 3. Diagrams of James Thomson's mechanical integrator

In 1876 William was searching for a mechanical way of obtaining the Fourier coefficients of a function. In discussions with his brother, James told him that he had designed a mechanical integrator some years earlier but had never published the details.

As Charlie Care recounts "In a flash of inspiration Kelvin saw how the mechanical integrator could offer 'a much simpler means of attaining my special object than anything I had been able to think of previously'." [15]

From this revelation, William moved with rapid speed. Within days, four influential papers were prepared to be given before the Royal Society of London. The first was written by James and described his integrator in detail; the others were by William and discussed their uses. These papers testify to the significance of the mechanical integrator, broadcasting to the world of science that it was possible to integrate products, solve second order differential equations, and with a particular set-up, solve differential equations of an arbitrary order.

It was not long before Kelvin's insight was incorporated into the harmonic analyser. In use, the harmonic analyzer derived the composite harmonics of tidal data and solved equations at the Meteorological Office. As a technology, it ushered in a new genre of calculating instrument - the continuous calculating machine.

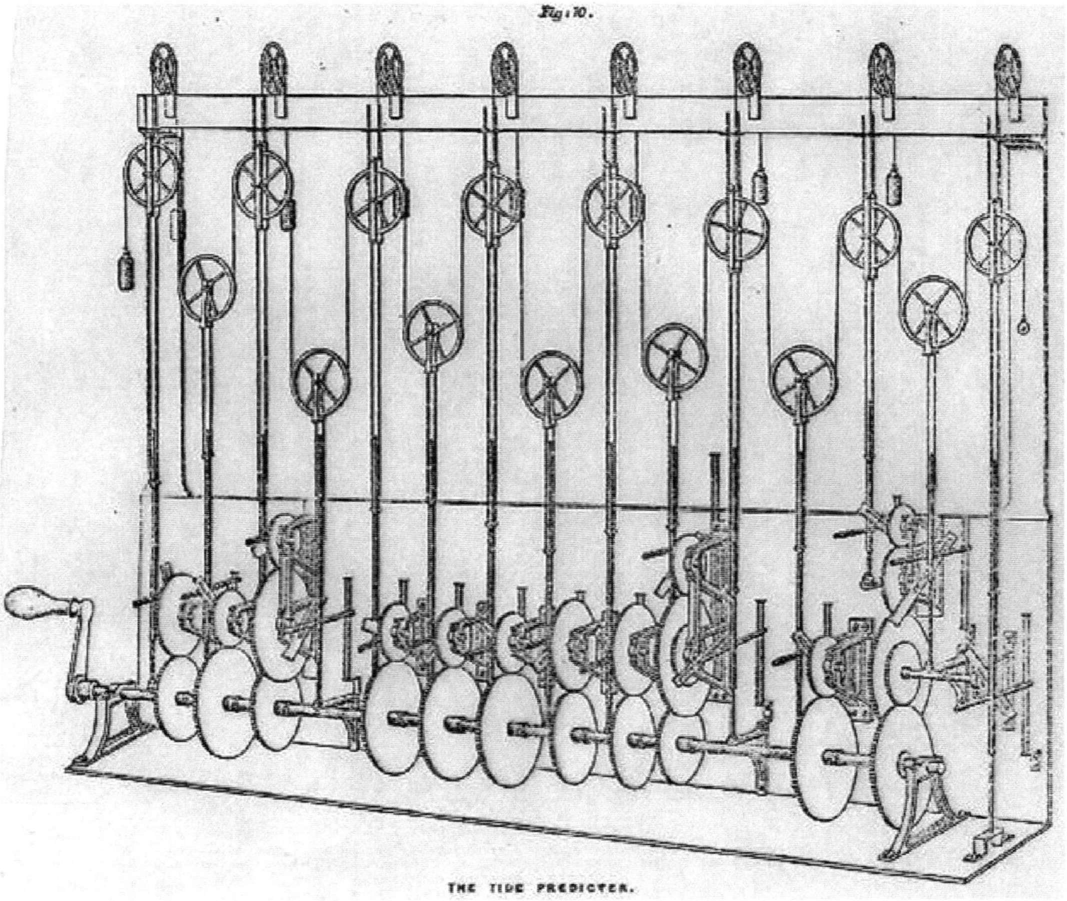

Figure 4. William Thomson's tide predictor

A discussion of continuous calculating machines was also in an appendix of the highly influential Treatise on Natural Philosophy by Thomson and Tait in its second edition of 1879. The Treatise was first published in 1867.and in it they aimed to re-centre natural philosophy around the principle of conservation of energy and the principle of least action.

5. Thomson and Tait

On October 29, 1902, William Thomson, now Lord Kelvin since his ennoblement in 1892, unveiled a portrait in Peterhouse College, Cambridge of his friend and collaborator Peter Guthrie Tait who had died in July of the previous year aged seventy. Although they had both been students at Peterhouse their times there did not overlap, Thomson being the elder of the two by some six years. The *Cambridge Chronicle* reported Kelvin, in his address at the unveiling, describing their friendship as beginning about 1860, when Tait came to Scotland from Queen's College, Belfast to succeed Forbes as Professor of Natural Philosophy at Edinburgh. The report said that Kelvin described their collaboration in these terms:

Journal of Physics: Conference Series **158** (2009) 012001 doi:10.1088/1742-6596/158/1/012001

> … he [Tait] was always ready with delightful quotations, and these brightened their hours of work. For they did heavy mathematical work, stone breaking was not in it. A propos, perhaps, of the agonies (he did not mean pains, he meant struggles) of the mathematical problems which they had always with them. [16]

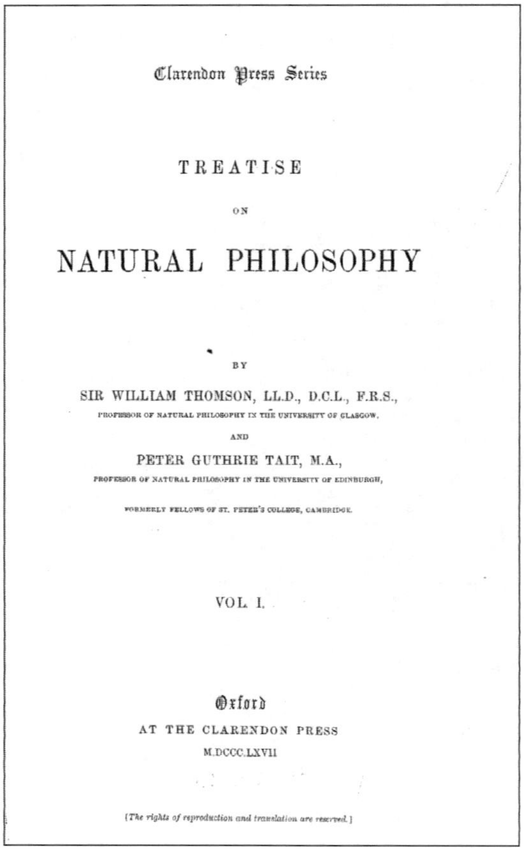
Clarendon Press Series

TREATISE

ON

NATURAL PHILOSOPHY

BY

SIR WILLIAM THOMSON, LL.D., D.C.L., F.R.S.,

PROFESSOR OF NATURAL PHILOSOPHY IN THE UNIVERSITY OF GLASGOW,

AND

PETER GUTHRIE TAIT, M.A.,

PROFESSOR OF NATURAL PHILOSOPHY IN THE UNIVERSITY OF EDINBURGH,

FORMERLY FELLOWS OF ST. PETER'S COLLEGE, CAMBRIDGE.

VOL. I.

Oxford

AT THE CLARENDON PRESS

M.DCCC.LXVII

[*The rights of reproduction and translation are reserved.*]

Figure 5. The work of 'two northern wizards', *The Treatise on Natural Philosophy* was a self conscious attempt to inherit the Newtonian crown.

Probably the most influential legacy of their collaboration on "heavy mathematical work" was the production in 1867 of *The Treatise on Natural Philosophy* by Thomson and Tait – still so called even after Kelvin's ennoblement. They started collaborating on it in 1861 shortly after they met and although it fell far short in scope of the original intentions of its authors it was to be highly influential in identifying and placing conservation of energy at the heart of its approach.

The two men were of very different natures. Thomson frequently went travelling while Tait did not leave Scotland after 1875. Tait could also be argumentative with bitter disputes for example with Heaviside and Gibbs over the relative merits of the vector approach as compared to using quaternions. However it was Tait who drove the collaboration towards publication – in turn cajoling, coaxing and berating Thomson to try to get him to keep to deadlines and schedules. Tait's frustration is illustrated in a letter he writes to Thomson in June 1864 about halfway through the collaboration where he says:

Journal of Physics: Conference Series **158** (2009) 012001 doi:10.1088/1742-6596/158/1/012001

> "I am getting quite sick of the great Book. . . if you send only scraps and these at rare intervals, what can I do? You have not given me even a hint as to what you want done in our present chapter about statics of liquids and gases! I have kinetics of a particle almost ready, nearly the whole of the next chapter, but I don't see the fun of paying 30/- for sending the MSS to you [in Germany] for revision, when in all probability you won't look at it till some indefinite period when you are in Arran, where it would be certain of reaching you — and for 8d. Now all this is very pitiable: I declare you did twice as much during the winter as you are doing now." [17]

The treatise was universally known as T & T′ and Thomson and Tait used the abbreviations themselves in their extensive correspondence. One of the formulae in the treatise was $d\rho / dt = jcm$ and as a result, their close friend James Clerk Maxwell became known as $d\rho / dt$! Maxwell and Tait had been friends since they first went to school at Edinburgh Academy at the age of 10, and from there to Edinburgh University and then onto Cambridge.

A LECTURE ON THOMSON'S GALVANOMETER

DELIVERED TO A SINGLE PUPIL IN AN ALCOVE WITH DRAWN CURTAINS

THE lamp-light falls on blackened walls,
And streams through narrow perforations;
The long beam trails o'er pasteboard scales,
With slow-decaying oscillations.
Flow, current! flow! set the quick light-spot flying!
Flow, current! answer, light-spot! flashing, quivering, dying.

O look! how queer! how thin and clear,
And thinner, clearer, sharper growing,
This gliding fire, with central wire
The fine degrees distinctly showing.
Swing, magnet! swing! advancing and receding;
Swing, magnet! answer, dearest, what's your final reading?

O love! you fail to read the scale
Correct to tenths of a division;
To mirror heaven those eyes were given,
And not for methods of precision.
Break, contact! break! set the free light-spot flying!
Break, contact! rest thee, magnet! swinging, creeping, dying.

$\frac{dp}{dt}$

Figure 6. James Clerk Maxwell waxing lyrical in the journal Nature, May 1872

Figure 7. The Reid Portrait of Peter Guthrie Tait at Peterhouse College, Cambridge

In 1852 Tait had graduated as senior Wrangler and Maxwell in 1854 as second Wrangler. After a couple of years in Cambridge Tait obtained the chair of mathematics at the Queen's College, Belfast where he worked closely with Thomas Andrews and William Thomson's brother James. It was during this period that he discovered Hamilton's work on quaternions for which he remained a lifelong enthusiast. Meanwhile Maxwell had become Professor of Natural Philosophy at Marischal College in Aberdeen but he had to leave as a result of the 1860 merger of Marischal with King's College because he was junior in the ranking.

Meanwhile JD Forbes relinquished the Chair of Natural Philosophy at Edinburgh University in 1859 and Maxwell and Tait both applied. Tait had nineteen testimonials including ones from Andrews, Hamilton, Thomson, Boole, Todhunter, and Challis. The deciding factor seems to have been teaching ability and on this basis Tait was appointed. Indeed the Edinburgh newspaper *The Courant* on reporting his election remarked

> ".... and we do not mean to dispute the decision of the curators, by saying, that in Professor Maxwell the curators would have had the opportunity of associating with the University one who is already acknowledged to be one of the remarkable men known to the scientific world. ... But there is another power which is desirable in a professor of a University with a system like ours, and that is, the power of oral exposition proceeding upon the supposition of a previous imperfect knowledge, of even total ignorance, of the study on the part of pupils." [18]

Tait was an enthusiastic, committed and lucid teacher - appreciated by colleagues and students alike. JM Barrie, more famous as the author of *Peter Pan* but previously an undergraduate at Edinburgh, paints a picture of Tait as a captivating and popular lecturer:

> "The small twinkling eyes had a fascinating gleam in them; he could concentrate them until they held the object looked at; when they flashed round the room he seemed to have drawn a rapier. I have seen a man fall back in alarm under Tait's eyes, though there were a dozen benches between them. These eyes could be merry as a boy's, though, as when he turned a tube of water on students who would insist on crowding too near an experiment." [19]

As I've mentioned Tait was an enthusiast for quaternions and first started corresponding with Hamilton while he was in Queen's and David Wilkins of Trinity has gathered together some of the correspondence between them during 1858 and 1859 while Tait was at Queen's and Hamilton at Trinity.

Towards the end of his life Kelvin was appointed Chancellor of Glasgow University and in his inaugural address on 29th November 1904 he reflects on his origins and influences:

> "To be Chancellor of one of the Universities of our country is indeed a distinguished honour. For me, to be Chancellor of this, my beloved University of Glasgow, is more than an honour. I am a child of the University of Glasgow. I have lived in it 67 years.
> But my veneration for the ancient Scottish University, then practically the University for Ulster, began earlier than that happy part of my life. My father, born in County Down, was for four years (1810 to 1814) a student of the University of Glasgow; and in his Irish home as first Professor of Mathematics in the newly-founded Royal Belfast Academical Institution, his children were taught to venerate the University of Glasgow. One of my earliest memories of those old Belfast days is of 1829, when the joyful intelligence came that the Senate of the University of Glasgow had conferred the honorary degree of Doctor of Laws on my father. Two years later came the announcement that the Faculty of Glasgow College had elected him to the Professorship of Mathematics. My father's experiences as a Glasgow student are naturally of supreme interest to myself. May I briefly speak of them, not because your kindness to me tempts me to be egotistical, but because the difficulties overcome and the precious life-long benefits won, by the struggles of an Irish student of the University of Glasgow in the beginning of the nineteenth century, illustrate the vitality and efficiency of the University in that primitive time. There were no steamers, nor railways, nor motor cars in those days. Can young persons of the present time imagine life to be possible under such conditions? My father and his comrade students, chiefly aspirants for the ministry

> of the Presbyterian Synod of Ulster and for the medical profession in the North of Ireland, had to cross the Channel twice a year in whatever sailing craft they could find to take them." [20]

So we see that Kelvin's view was that there was a close association between Ulster and Glasgow University which brought great advantages to both of them. His Ulster roots and family background informed his attitudes, personality and priorities and had laid a foundation for his subsequent life and labours at Glasgow.

LORD AND LADY KELVIN IN CORONATION DRESS

Figure 8. The man of Empire. Lord and Lady Kelvin dressed for the coronation of King Edward VII on August 9th 1902.

Journal of Physics: Conference Series **158** (2009) 012001 doi:10.1088/1742-6596/158/1/012001

References

[1] Flood R, McCartney M and Whitaker MAB 2008 *Kelvin: Life, Labours and Legacy* (Oxford: Oxford University Press)

[2] Thomson J 1837 *Treatise on Arithmetic in Theory and Practise* (18/e) (Belfast: Simms & McIntyre)

[3] King ET (ed) 1909 *Lord Kelvin's Early Home* (London: Macmillan & Co Ltd) p126

[4] King ET (ed) 1909 *Lord Kelvin's Early Home* (London: Macmillan & Co Ltd) p41

[5] Thompson SP 1910 *The Life of William Thomson, Baron Kelvin of Largs* (London: Macmillan & Co Ltd) p17

[6] Thompson SP 1910 *The Life of William Thomson, Baron Kelvin of Largs* (London: Macmillan & Co Ltd) p32

[7] Thompson SP 1910 *The Life of William Thomson, Baron Kelvin of Largs* (London: Macmillan & Co Ltd) p37

[8] King ET (ed) 1909 *Lord Kelvin's Early Home* (London: Macmillan & Co Ltd) p204

[9] King AG 1925 *Kelvin the Man* (London: Hodder and Stoughton Ltd) p87

[10] Thompson SP 1910 *The Life of William Thomson, Baron Kelvin of Largs* (London: Macmillan & Co Ltd) p87

[11] Thompson SP 1910 *The Life of William Thomson, Baron Kelvin of Largs* (London: Macmillan & Co Ltd) p48

[12] Thompson SP 1910 *The Life of William Thomson, Baron Kelvin of Largs* (London: Macmillan & Co Ltd) p1154

[13] Sharlin HI 1979 *Lord Kelvin: The Dynamic Victorian* (Pennnsylvania & London: Pennsylvania State Univeristy Press) p73

[14] Bowler P 2008 James Thomson and the Culture of a Victorian Engineer *Kelvin: Life, Labours and Legacy* ed Flood R, McCartney M and Whitaker MAB (Oxford: Oxford University Press) p57

[15] www.rutherfordjournal.org/article020106.html [last accessed August 2008]

[16] Knott CG 1911 *Life and Scientific Work of Peter Guthrie Tait* (Cambridge: Cambridge University Press) p50

[17] Smith C and Wise MN 1989 *Energy anf Empire: A Biographical Study of Lord Kelvin* (Cambridge: Cambridge University Press) p450

[18] Knott CG 1911 *Life and Scientific Work of Peter Guthrie Tait* (Cambridge: Cambridge University Press) p16

[19] Barrie JM 1894 *An Edinburgh Eleven* (London: Hodder and Stoughton Ltd) p46-47

[20] Thomson W 1911 Address as Chancellor at Glasgow *Mathematical and Physical Papers Volume 6* (Cambridge: Cambridge Univerisity Press) p370-371

From a land of saints and scholars? Reflections on William Thomson's religious background

Mark McCartney

School of Computing and Mathematics, University of Ulster, Newtownabbey, Co. Antrim, Northern Ireland, BT37 0QB

m.mccartney@ulster.ac.uk

Abstract. After indicating the range of biographical studies of William Thomson since his death in 1907, the roots of Thomson's religious views are traced back to those of his father, James Thomson, and following the work of David B Wilson, to those of his teachers at the University of Glasgow in the 1830s.

1. The many lives of William Thomson

William Thomson was, in his lifetime, perhaps the most famous physical scientist in Britain. His range of scientific contributions was wide, encompassing electricity and magnetism, the age of the earth, thermodynamics, and the nature of the ether. As an engineer he worked on the laying of the first transatlantic telegraph cable, the development of electrical instruments, the setting of standards for electrical units and the improvement and development of a range of maritime devices. His scientific output was considerable, with the list of publications given at the end of Silvanus P Thompson's official biography running to 661 entries [1].

As JJ Thomson said during the 1924 celebrations organised to commemorate the centenary of Thomson's birth "To be interested in a thing was, with Lord Kelvin, synonymous with discovering something new about it" [2]. And one might add in certain cases 'and with making money out of it'. The quip is not intended to be derogatory, merely a statement of fact. William Thomson took out 70 patents in his lifetime [3] and his work on cable laying, and maritime and electrical instruments made him a wealthy man, resulting in a fine 126 ton yacht, an impressive home west of Glasgow, and an estimate of his worth at time of death of £169,000 [4].

Table 1 gives an overview of Thomson's life, and in the impressive list of achievements and awards two pivots can be placed. The first is his reading of Fourier's *Théorie Analytique de la Chaleur* in May 1840. This is noteworthy, not just because it shows high mathematical ability, but also because Fourier's work was a constant intellectual companion and source of inspiration to him throughout his life. Thus to quote JJ Thomson again "… I have heard him [i.e. William Thomson] say …that whenever he had done anything with which he felt particularly pleased, Fourier's theorem was always at the bottom of it."[5] SP Thompson states: "About certain branches of mathematics he was always enthusiastic. In praise of

Fourier's "mathematical poem," *Théorie Analytique de la Chaleur,* he never tired" [6]. Or finally there is PG Tait's blunt and direct assessment "Fourier made Thomson" [7]. Fourier's

Table 1. A brief chronology of William Thomson's life

Date	Event
26th June 1824	William Thomson born in Belfast
October 1834	Matriculates at Glasgow (aged 10)
May 1840	Reads Fourier's *Théorie Analytique de la Chaleur* in a fortnight (aged 15)
October 1840	Enters St Peter's College, Cambridge (aged 16)
May 1841	First paper published in the Cambridge Mathematical Journal (aged 16)
January 1845	Second wrangler and first Smith's Prizeman (aged 20)
June 1845	Elected Fellow of St Peter's College, Cambridge (aged 21)
September 1846	Unanimously elected to Chair of Natural Philosophy in Glasgow (aged 22)
February 1847	Elected Fellow of the Royal Society of Edinburgh. Served as President 1873-8, 1886-90, 1895-1907. RSE was one of the first of over 100 learned societies across the world to elect Thomson to their ranks.
June 1851	Elected Fellow of the Royal Society of London (aged 26). Served as PRS 1890-5.
1857	Honorary Doctor of Laws, Dublin University. This is the first of 21 honorary doctorates which Thomson was awarded.
November 1866	Knighted by Queen Victoria (aged 42)
January 1892	Elevated to Peerage
June 1902	Awarded the Order of Merit
August 1902	Made Privy Councillor
April 1904	Elected Chancellor of Glasgow University
17th December 1907	Dies. Buried beside Newton in Westminster Abbey on 23rd.

methods formed the basis of Thomson's first published paper in which he, as a mere 16 year old, defended Fourier against the erroneous criticisms of Edinburgh University's Professor of Mathematics Philip Kelland. Then throughout his career he went on to employ Fourier in work on the cooling of the earth, telegraphy and tide prediction. Tait's waspish remark might be more accurately rendered 'Thomson made very good use of Fourier'.

The second pivotal event in table 1 is Thomson's knighthood in November 1866 – this may be seen as signaling the apex of his career. The knighthood resulted directly from Thomson's heavy involvement in the laying of the transatlantic telegraph cable. The success of the cable (after repeated failures) resulted in a huge change in how the two continents could communicate and Thomson was at the very centre of the project. During work before the actual cable laying the Glasgow to London train would be delayed to ensure Thomson's travel [8], and when the cable was finally laid *The Times* announced it as "what deserves to be regarded as in many respects the most wonderful achievement of this victorious century" [9]. Thomson was a national figure. His work on the cable formed fertile ground for his science and entrepreneurship. Not only were there the direct results of papers on electricity, the development of new equipment (such as the mirror galvanometer) and patents and wealth, but Thomson also got a taste for life at sea. This resulted in the previously mentioned purchase of a yacht and a whole series of papers and patents on maritime issues. Thus in the 1870s and 1880s he gave over 50 papers and talks on matters such as the tides, compass design, lighthouse signaling and depth sounding.

Figure 1. Dustcover of the 1938 biography of Kelvin by Alexander Russell. Other volumes in the 'Order of Merit' series included biographies of JM Barrie, Sir Edward Elgar and Thomas Hardy.

As a subject for biographical study Thomson has been perennially popular. Within months of his death his successor in the Chair of Natural Philosophy at Glasgow, Andrew Gray, published a 316 page biography [10], and that was followed the next year by a book of reminiscences of Thomson's childhood collected together by his niece [11]. The year 1910 saw the publication of the official biography by Silvanus P Thompson [12] – a two volume work which remains a readable mine of information to this day. There are also other shorter pieces dated around this time– see for example [13-16], but it is not until 1924, the centenary of his birth, that we find further single volumes devoted to Thomson's life, with a biography by another niece [17], a short sketch of his early career at Glasgow [18] and a book resulting from the official centenary celebrations in London [19]. This last book contains not only an essay by Sir JJ Thomson, but a remarkable collection of some 29 "Addresses Commemorative" from a range learned societies and academies writing of Thomson in brief and glowing terms (see for example figure 2).

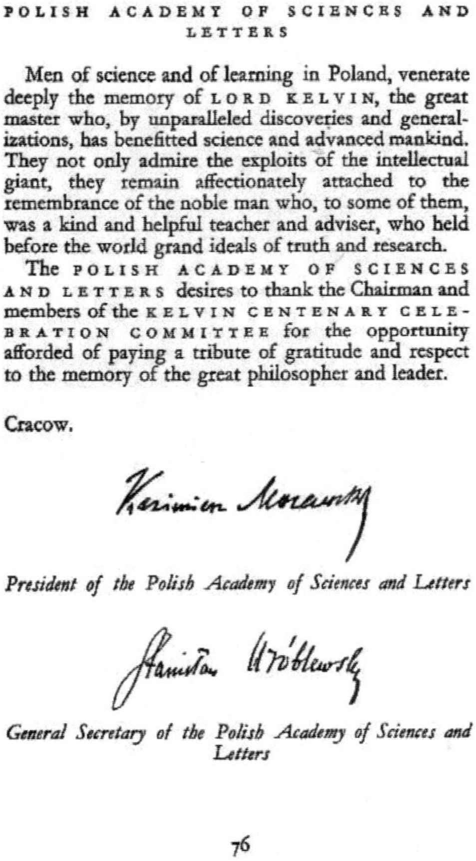

POLISH ACADEMY OF SCIENCES AND LETTERS

Men of science and of learning in Poland, venerate deeply the memory of LORD KELVIN, the great master who, by unparalleled discoveries and generalizations, has benefitted science and advanced mankind. They not only admire the exploits of the intellectual giant, they remain affectionately attached to the remembrance of the noble man who, to some of them, was a kind and helpful teacher and adviser, who held before the world grand ideals of truth and research.

The POLISH ACADEMY OF SCIENCES AND LETTERS desires to thank the Chairman and members of the KELVIN CENTENARY CELEBRATION COMMITTEE for the opportunity afforded of paying a tribute of gratitude and respect to the memory of the great philosopher and leader.

Cracow.

President of the Polish Academy of Sciences and Letters

General Secretary of the Polish Academy of Sciences and Letters

76

Figure 2. Venerating the good Lord Kelvin. One of the many 'Commemorative Addresses' sent from learned societies from across the globe as part of the celebrations to commemorate the centenary of William Thomson's birth in 1924.

It is not until 1938 that another full length biography is published – this time in a series of books on members of the Order of Merit [20]. From 1938, with the exception of two briefer works [21, 22] it seems we have to wait until 1974 and the publication of Joe Burchfield's *Lord Kelvin and the Age of the Earth* [23] to find a revival of interest (at least in book form) in Thomson. After Burchfield's book comes a steady stream of others [24-29] including the publication of the Kelvin-Stokes correspondence [30]. Of these more recent books undoubtedly the most encyclopaedic is *Energy and Empire* by Crosbie Smith and M Norton Wise [26], while a number of the others emphasize the importance of particular facets of his life. Thus, to put it in crude terms, by sheer bibliographic weight alone the importance of Thomson is emphasised. But his life begins in no less a place than Belfast, and his paths were carefully directed by his father.

2. A barbarous nook in Ireland

Ireland as a whole, and more recently the north of Ireland in particular, has a persistent reputation for political and religious strife which sits uneasily with its medieval lore of being a 'land of saints and scholars'. William's father, James Thomson, saw the outworking of such strife both in his childhood, and early working life.

James Thomson's great-great-grandfather had arrived to settle near Ballynahinch in Co. Down around 1641 [31], at the tail end of the Ulster Plantation. Thus perhaps he witnessed aspects of the Rebellion of 1641, when the native (Roman Catholic) Irish rose up in attempt to repossess land given to the incoming (Protestant) planters.

Winding forward more than one hundred and fifty years, we can be sure that as a boy James Thomson witnessed part of another uprising – the Rebellion of 1798 – because in later life he wrote about one of the clashes between crown forces and the rebels. It occurred in the town of his birth and became known as the Battle of Ballynahinch. He helped carry provisions to the rebel camp before the battle, and then, returning home, watched from a nearby hill as the King's forces approached in a pincher movement from Belfast (in the north) and Downpatrick (the south). Writing in 1825 James Thomson recalls "the approach of the party from Belfast was in a short time announced, by the smoke and flames of the farm-houses, which they set on fire indiscriminately, on their march from Saintfield to Ballynahinch." The fighting was engaged for approximately three hours from about 6pm on the evening on Tuesday 12th of June 1798 and

> "during this period the advantage lay on the side of the military, in consequence of their superior discipline and appointments." In the early hours of Wednesday morning "the horrors of the scene were renewed, by the King's forces setting fire to the town; and in a short time, a great proportion of the best houses in it were enveloped in flames, and hastened to inevitable destruction. This act, which was by no means necessary, caused the rebels immediately to recommence the fight, and to endeavor, by means of their small artillery, to arrest the work of devastation." [32]

It was more of a rout than a battle and the rebels, who were armed for the most part only with pikes, and who deserted heavily overnight, were no match for crown forces.

A crucial difference between the Rebellion of 1641 and that of 1798 was that it was not inherently religiously sectarian. The rebels of 1798 were the United Irishmen – a movement founded in Belfast in 1791 by a small group of Presbyterians, but which quickly grew across the island to be made up of Roman Catholics, Presbyterians and other dissenters, and a small number of Anglicans. They took their inspiration from the French and American revolutions. Their aims were independence for Ireland, parliamentary reform, and equality of rights in a land where the members of the established Church of Ireland held an overwhelming control. When the rebellion rose they were, as exampled by the quote above, decisively and ruthlessly defeated across Ireland. The short lived ideological union between Protestant and Roman Catholic was demolished and has never reappeared in the north of Ireland.

By 1814 Belfast could no longer be described by John Milton's acerbic words as "a barbarous nook in Ireland" [33]. Its population and prosperity were growing, and its desire for intellectual growth was signaled by the founding of a new seat of learning for the youth of the town- Belfast Academical Institution (locally known simply as Belfast Inst). One of the founders was Dr William Drennan, himself a onetime United Irishman, who stated at the opening ceremony of the college "pupils of all religious denominations should communicate… in the common business of education, by which means a new turn might be given to the national character and habits."[34] The existence of men like Drennan, and the formation of Belfast Inst are indicative of a significant strand of political radicalism and religious liberalism within Belfast at the start of the nineteenth century.

James Thomson was employed in the Belfast Inst as its first teacher of mathematics. He was most certainly in sympathy with the educational sentiments expressed by Drennan, and the two men became close friends. A love for learning and the value of hard work were all forged in James Thomson's childhood. Indeed as a child he taught himself arithmetic and, with only one example of a sundial to help him, constructed his own sun and night dials. As an adult he would rise at 4 am to work on his many successful textbooks (which covered topics such as arithmetic, calculus, trigonometry and geography) and devoted himself tirelessly to his own children's education. James Thomson's political views must surely have been influenced by what he saw of the Battle of Ballynahinch, but this was merely the confirming of the pro- United Irishmen stance of his childhood home. Indeed James Thomson's daughter notes of her father that "His sisters taught him to read, using for this purpose handkerchiefs on which were printed mottoes and verses composed by the patriots who brought about the rebellion of 1798." [35] Years later, when his own children were growing up they would imagine that the chimney was populated by wicked "Wee Tories", and one child (John) once announced to a (Tory) dinner guest "I'm a radical reformer." [36] James Thomson certainly passed his political views onto his children!

Belfast Inst was not just a school for the young boys of Belfast however. It was a school and college combined, with the college offering an education aimed to be comparable to an MA from a Scottish university. Thus it would enable Presbyterian ministers to be educated locally, rather than having to make the, previously common, journey to Glasgow University. Note that 'Presbyterian' is here being used as a blanket term covering the General Synod of Ulster, and Seceders, the two major (but independent) groupings of Presbyterians within the north of Ireland. Belfast Inst attempted to maintain its non-sectarian ideals by asking the churches to appoint their own divinity professors, with the appointees then being given space in the Institution for their classes, and seats on the faculty. But all was not well. Those within the conservative wing of the Synod of Ulster were unhappy with men like Drennan. Drennan was politically radical, pro-catholic and the son of a 'New Light' Presbyterian minister (a term that indicated "the less orthodox, and often the more urbane and erudite, body of Ulster Presbyterian" [37]).

Vocal in his criticisms of Belfast Inst was Henry Cooke, a Presbyterian minister who claimed that it was a place of learning that was corrupted by staff who were Arians (i.e. denied the divinity of Christ) and thus were by their influence going to undermine the orthodoxy of future ministers. Cooke's opposition to the Institution, and his political conservatism did nothing to endear him to James Thomson in his early years in Belfast. However, although Thomson may have had sympathies with Drennan's radical politics, he did not share his New Light theology.

Thomson's Christian faith was deep and orthodox, but not outspoken. Thus, though he may have disliked Cooke's politics and stance towards Belfast Inst he was probably, on purely doctrinal issues, not that far from him. Or as one modern Irish Presbyterian quipped "Henry Cooke['s] theology was sounder that his politics" [38]. This is perhaps illustrated by Thomson's attitude to the evangelical Scottish theologian and academic, Thomas Chalmers. Thomson's warmth of feeling towards Chalmers can be measured by the fact that Chalmers stayed for a week with James Thomson and his family on a visit to Belfast in 1827, baptising his youngest daughter Margaret and opening the church where the family worshipped. Yet it was Chalmers who, sixteen years later, led the so called Disruption of the Church of Scotland in 1843. The Disruption split the denomination, with Chalmers and his supporters forming the Free Church of Scotland. Further, although in his sermon at the opening of the Thomsons' place of worship Chalmers made comments which were perceived to be critical of Cooke's stance, (so much so that Cooke wrote to him to ask for clarification) Chalmers wrote in his diary that Cooke was the "most impressive" minister he had met [39]. Thus we see that though James Thomson certainly disliked Cooke, which on a naive reading might lead us to classify him as theologically opposed to him, his relationship with Chalmers would seem to imply a rigorous theological orthodoxy. Further as Andrew R Holmes notes [40], Thomson's old teacher from Ballynahinch, and his minister in Belfast were both on the conservative wing of Presbyterianism. Both also ended up as his colleagues at Belfast Inst – one as divinity professor for the Seceders, the other for the Synod of Ulster.

Indeed at a superficial level it is almost ironic that Thomson, who wrote to his son William at Cambridge stating "Recollect my invaluable maxim never to quarrel with a man (but to waive the subject) about religion" [41] should admire Chalmers, but dislike Cooke. For whereas Chalmers split the Church of Scotland, Cooke, according to one modern Irish Presbyterian, by his very zeal for orthodoxy "paved the way for the eventual union of the Secession Synod with the Synod of Ulster to form the General Assembly of the Presbyterian Church in Ireland in 1840 [i.e. the Irish Presbyterian church in its modern form]." [42]

3. A Cherished Strong Religious Faith

The Concise Dictionary of National Biography [43] concludes its entry on William Thomson by stating that he "cherished through life a strong religious faith." This, as we shall see, was certainly manifested in his adult life. However as a child, his fulsome sense of humour concealed any future depth of faith from at least one Church of Scotland minister: William's older sister, Elizabeth, relates the story of a summer holiday in 1834 on the Scottish Island of Arran. The Thomsons worshipped in Lamlash Parish Church (figure 3) and

> "It happened one Sunday that there was a revival service, and the congregation grew much excited, uttering loud exclamations and groans; and at last some of the old women began to give vent to their feelings by tossing their Bibles in the air. This tickled Willie's sense of humour, and he shook with smothered laughter,

> which started all the other boys laughing too. Our pew was close under the pulpit in full view of the preacher, who, looking down, administered a grave rebuke. The smothered laughter then exploded, and the minister, pointing his finger at the ringleader, exclaimed 'Ye'll no lach when ye're in hell!' This was too much: and Willie rolled clean over on the floor. For some reason or other our father was not with us that day, and I was in charge of the party. Crimson with shame, I bustled them all out of the church as quickly as I could." [44]

Figure 3. The old Parish Church, Lamlash, Island of Arran (demolished circa. 1884). The boarded up windows probably indicate that the picture was taken just prior to demolition. (Courtesy of Capt. JL Davidson).

This is, of course, simply an example of childhood levity, but in adulthood we find evidence enough of piety. Thus at the end of his life his niece records how, on a visit to see her and her sister in their new home "as he was stepping into the carriage to leave, he turned and sweetly and reverently blessed our home and blessed us." [45]. On a Sunday morning in June 1872, aboard his yacht at Gravesend, east of London "...just as they weigh anchor the "Thames Mission" boat comes up, and Sir William orders Captain Flarty to stop the yacht while the minister conducts a service for them on board." [46] The journey he was embarking on took him on a tour of telegraph stations in Gibraltar, Lisbon and Porthcurno, and in a letter to his sister-in-law, after commenting on the fact that he had to carry out tests on the lines in Gibraltar on a Sunday, he adds in a footnote that such work was "lawful on the ground of necessity and mercy" [47]. The phrase "necessity and mercy" comes directly from the Shorter Catechism – a short book of questions and answers about Christian doctrine which Thomson, as a child brought up in the Presbyterian church, may well have learnt by rote. Question 60: "How is the Sabbath to be sanctified?" gives the answer,

> "The Sabbath is to be sanctified, by a holy resting on that day, even from such worldly employments and recreations as are lawful on other days; and spending the whole time in the public and private exercises of God's worship, except so much as is to be taken up in the works of necessity and mercy."

In the course of his life Thomson certainly appears to have put full emphasis on this last clause in the answer and to downplay the earlier ones. Thus, he was happy to work on mathematics on a Sunday, or that museums, galleries and gardens be opened to the public - views which would have been anathema to Sabbatarians.

One final example of William Thomson's faith is given by his niece, Agnes Gardiner King:

> "When he was staying with us in London in 1888, Sunday proved too wet for us to venture to church, and he suggested our having a Bible-reading together. To quote from my sister, Elizabeth Thomson King's diary: 'Bibles were brought out, old version and new, and he began turning up passages, discussing the genuineness of Old Testament history, comparing dates about prophecies and consulting Biblical dictionaries in relation to debated points, all with the liveliest interest...He remarked what a pity it is that maps are not more universally given in Bibles, as it would add so very much to the interest of the study of the book; he also thought it was a great blank not having the dates at the top of the pages in the new version. Where they are only guessed at in the very ancient records he would have them put with a question mark, but when we get to the time of Solomon and so on the dates are as accurately known, he says, as that of the battle of Hastings. In the evening mother read aloud Darwin's confession of faith, or rather his confession of want of faith. Uncle William said of his views about the absence of evidence of design, that he considered such views utterly unscientific. He expressed himself very vehemently on the subject, and said that our own power of discussing and speculating about Atheism and Materialism was enough of itself to disprove such a theory. With regard to Evolution, he said that it could not in the least degree explain the great mystery of nature and creation: if all things originated in a single germ, then that germ contained in it all the marvels of creation, physical, intellectual and spiritual to be afterwards developed. It is absolutely impossible that atoms of dead matter should come together so as to make life.'" [48]

The examples given above all illustrate William Thomson's private faith. But it also appeared in his working life. It most frequently broke into the public domain with his use of the third Collect for Grace from the Book of Common Prayer at the beginning of his classes at Glasgow University. It is not that he prayed at the beginning of his classes that is significant, that would have been standard practice at Glasgow. Rather it was the fact that Thomson chose to use a prayer from the *Anglican* prayer book to commence classes in a university, which up until 1850s formally required its professors to sign the University tests. The University tests were the formal doctrinal statements of the (Presbyterian) Church of Scotland (the Westminster Confession of Faith) and a document declaring intention to conform to Church of Scotland worship and church discipline on appointment to their chairs. This example of William Thomson's broad churchmanship was inherited directly from his father, who as Professor of Mathematics at Glasgow, had since the early 1840s been one of the agitators for the removal for of the requirement to sign the tests. Neither was its isolated – as is noted by Iain Hutchison [49] he attended Anglican, Church of Scotland and Free Church of Scotland freely in an era when ecclesiastical, and related doctrinal, fidelity were taken seriously. One example of such a serious approach in others can be seen in the fact that GG Stokes demurred from Thomson's requests that he apply for the Chair of Mathematics at Glasgow on the grounds that he was an Anglican, and was not willing to become a Presbyterian. Thus, although willing to profess his "non-hostility to the established church of Scotland" he could not sign the University tests as "[t]here is no doubt that the test means, & was intended in the first instance to mean, that the subscriber is a bona fide member of the established Church of Scotland."[50]

4. The Great First Cause

David Wilson traces the roots of William Thomson's links between science and faith to his education at Glasgow and Cambridge. For the influences at Cambridge, Wilson notes the writings of William Whewell and John Herschel, and the views of Thomson's coach for the Tripos, William Hopkins. But more striking are the influences at Glasgow, principally because some of these come directly from the lecture notes Thomson took while an undergraduate. Thus from notes taken in Professor William Fleming's moral philosophy class: "Whether we look to the heaven above us, or the earth below us, we see striking marks of intelligence and design. And when we examine them particularly we see still more striking proofs." Or "Mind is as full of marks of design as matter." [51] From his notes from Professor William Meikleham's natural philosophy lectures; the subject under study not only extended man's dominion over nature and was intellectually satisfying but "above all, it leads us to view the Creator as the Great First Cause, and as maintaining the energies of nature."[51] Or from the work of the Professor of Astronomy, John Pringle Nichol (who due to Meikleham's ill health ended up teaching Thomson most of his natural philosophy at Glasgow); "it does appear that such precision of workmanship and stedfast solemnity of march [of the solar system], are…strong and eloquent proofs of the presence of the Godhead." [52]

We have in these quotes three strands. First, evidence of God's existence from design in nature, second evidence for God's existence from the existence of the human mind (and its capability to reason) and finally the necessity of God as the First Cause in creation. We note *en passant* that all three of these strands are present in Thomson's private discussions with his family as recorded by his niece in the previous section. But, as shown below, we also find them in his published work.

The evidences for design were most fully worked out by William Paley, who sought and saw such evidence throughout creation, variously in the human skeleton, muscle structure, the eye, comparative anatomy, the insect and plant kingdoms and astronomy [53]. Thomson, in his 1871 BA presidential address stated that "I feel profoundly convinced that the argument of design has been greatly too much lost sight of in recent zoological speculations" and, referring to Paley's *Natural Theology* commended

> "the solid and irrefragable argument so well put forward in that excellent old book…overpoweringly strong proofs of intelligent and benevolent design lie all around us, and if ever perplexities, whether metaphysical or scientific, turn us away from them for a time, they come back upon us with irresistible force, showing to us through nature the influence of a free will, and teaching us that all living beings depend on one ever-acting Creator and Ruler." [54]

Indeed in later life when Thomson was asked to give the Gifford lectures he declined, feeling he had nothing to add that was not already in Paley, and stating in a letter to PG Tait that "All I could say on the subject can be said in five minutes and I have already said it" [55].

Elsewhere in the same 1871 presidential address Thomson states that

> "Hence when the Earth was first fit for life, there was no living thing on it. There were rocks solid and disintegrated, water, air all round, warmed and illuminated by a brilliant Sun, ready to become a garden. Did grass and trees and flowers spring into existence, in all the fullness of ripe beauty, by a fiat of Creative Power? Or did vegetation, growing up from seed sown, spread and multiply over the whole Earth? Science is bound by the everlasting law of honour, to face fearlessly every problem which can fairly be presented to it. If a probable solution, consistent with the ordinary course of nature, can be found, we must not invoke an abnormal act of Creative Power." [56]

That said Thomson went on to add that if science were to provide an explanation for the arrival of life on earth, arrival by meteor was a possibility. Thus he merely, and pointedly, pushed the problem of the creation of life back one step: "Dead matter cannot become living matter without coming under the influence of matter previously alive." [57]

After quoting Darwin's concluding remarks in the *Origin of Species* which speak positively of a creator who initially breathes life into a "few forms or into one" he continues with the barbed,

> "With the feelings expressed in these two sentences I most cordially sympathise. I have omitted two sentences which come between them, describing briefly the hypothesis of "the origin of species by natural selection," because I have always felt that this hypothesis does not contain the true theory of evolution, if evolution there has been in biology." [58]

Note the last, and tellingly sceptical, phrase "if evolution there has been in biology."

In an article on the dissipation of energy in 1892 he writes

> "The considerations of ideal reversibility, by which Carnot was led to his theory, and the true reversibility of every motion in pure dynamics have no place in the world of life. Even to think of it (and on the merely dynamical hypothesis of life we can think of it as understandingly as of the origination of life and evolution of living beings without creative power), we must imagine men, with conscious knowledge of the future but with no memory of the past, growing backward and becoming again unborn; and plants growing downwards into the seeds from which they sprang." [59]

Thus for Thomson the idea life emerging and evolving without a creator, is about as likely as time running backwards.

In an address to the Glasgow Geological Society in 1868 he states

> "...all things are possible to Creative Power. But we know also, that Creative Power has created in our minds a wish to investigate and a capacity for investigating; and there is nothing too rash, there is nothing audacious, in questioning human assumptions regarding Creative Power." [60]

The above quotes from Thomson's writings indicate his confidence in the argument from design in nature, and an allied scepticism towards biological evolution. He emphasises the necessary intervention of a 'creative power' to generate life in the first place and notes that this creative power has also created the human mind and its capacity for inquiry. As noted at the beginning of this section, Wilson finds all of these rooted in Thomson's experience as an undergraduate.

5. Concluding remarks

We find the roots of William Thomson's religious views both in his father's deep faith and in the classes of those who taught him at Glasgow University. One approach would be to dismiss Thomson's faith on the grounds that it was purely an artefact of his early life from which he never escaped, or a function of the age in which he lived. I would view such an approach as being as useful as (for example) a dismissal of Thomson's use of Fourier's techniques on the grounds that he lived in the nineteenth century and that he first read Fourier's *Analytic Theory of Heat* as a teenager.

In conclusion rather, I would quote David Wilson's circumspect remarks:

> "No matter how great a man's genius may be, his ideas will be constrained, to an important extent, by the time and place in which he lives. Indeed in dealing with a man like Kelvin, it is incumbent upon us to spell out the relationship between the man and his surroundings. In Kelvin's case his theological views were not original, but they, nevertheless, constituted a fundamental component of his concept of nature." [61]

Figure 4. Kelvin in lofty company. Busts of Kevin, Roger Bacon, Newton and Archimedes are at the top of the Pillar of Science in St. Anne's Cathedral, Belfast. The pillar was erected in the memory of Thomas Andrews, sometime Professor of Chemistry and Vice Principal of Queen's College, Belfast. Andrews had been a colleague and collaborator of Kelvin's brother James Thomson during his time at Queen's. (Courtesy of SK Houston.)

Acknowledgment

I would like to thank Andrew R Holmes for taking the time to discuss aspects of Irish history (ecclesiastical and secular) with me. However, I am of course solely responsible for any errors of fact or interpretation.

References

[1] Thompson SP 1910 *The Life of William Thomson, Baron Kelvin of Largs* (London: Macmillan & Co Ltd) p1223
[2] Thomson JJ 1924 *Kelvin Centenary Oration and Addresses Commemorative* (London: Percy Lund Humphries & Co Ltd) p30
[3] Thompson SP 1910 *The Life of William Thomson, Baron Kelvin of Largs* (London: Macmillan & Co Ltd) p1275
[4] Smith C 2004 *Thomson, William, Baron Kelvin (1824–1907)* Oxford Dictionary of National Biography (Oxford: Oxford University Press) [http://www.oxforddnb.com/view/article/36507 last accessed August 2008]
[5] Thomson JJ 1924 *Kelvin Centenary Oration and Addresses Commemorative* (London: Percy Lund Humphries & Co Ltd) p10
[6] Thompson SP 1910 *The Life of William Thomson, Baron Kelvin of Largs* (London: Macmillan & Co Ltd) p1139
[7] Smith C and Wise MN 1989 *Energy and Empire: A biographical study of Lord Kelvin* (Cambridge: Cambridge University Press) p149
[8] Thompson SP 1910 *The Life of William Thomson, Baron Kelvin of Largs* (London: Macmillan & Co Ltd) p489
[9] Marchant J 1937 *History Through 'The Times': A Collection of Leading Articles on Important Events, 1800-1937* (Cassell and Company Ltd) p207
[10] Gray A 1908 *Lord Kelvin: An account of his scientific life and work* (London: JM Dent & Co)
[11] King ET (ed) 1909 *Lord Kelvin's Early Home* (London: Macmillan & Co Ltd) p126
[12] Thompson SP 1910 *The Life of William Thomson, Baron Kelvin of Largs* (London: Macmillan & Co Ltd) pp1297
[13] Griffiths AB 1912 *Biographies of Scientific Men* (London: Robert Sutton) p185-202
[14] MacNair P and Mort F (ed) 1908 *History of the Geological Society of Glasgow 1858-1908* (Glasgow: Glasgow Geological Society) p197-223
[15] Ramsay W 1908 *Essays Biographical and Chemical* (New York: EP Dutton & Co) p89-100
[16] Wilson D 1910 *William Thomson Lord Kelvin: His way of teaching natural philosophy* (Glasgow: John Smith & Co) pp56
[17] King AG 1924 *Kelvin the Man* (London: Hodder and Stoughton) pp142
[18] Murray D 1924 *Lord Kelvin: As professor in the old college of Glasgow* (Glasgow: Maclehose Jackson & Co) pp22
[19] Glazebrook R 1924 *Kelvin Centenary Oration and Addresses Commemorative* (London: Percy Lund Humphries & Co Ltd) pp100
[20] Russell A 1938 *Lord Kelvin* (Glasgow: Blackie & Son Ltd) pp163
[21] Young AP 1948 *Lord Kelvin: Physicist Mathematician Engineer* (London: British Council) 41pp
[22] MacDonald DKC 1965 *Faraday Maxwell and Kelvin* (London: Heinemann) pp143
[23] Burchfield JD 1975 *Lord Kelvin and the Age of the Earth* (New York: Science History Publications) pp260
[24] Sharlin HI 1979 *Lord Kelvin: The Dynamic Victorian* (Pennsylvania & London: Pennsylvania State University Press) pp272
[25] Wilson DB 1987 *Kelvin and Stokes: A Comparative Study in Victorian Physics* (Bristol: Adam Hilger) pp253

[26] Smith C and Wise MN 1989 *Energy and Empire: A biographical study of Lord Kelvin* (Cambridge: Cambridge University Press) pp866
[27] Tunbridge P 1992 *Lord Kelvin: his influence on electrical measurements and units* (London: Peter Peregrinus Ltd) pp107
[28] Lindley D 2004 *Degrees Kelvin: The Genius and Tragedy of William Thomson* (London: Aurum Press) pp256
[29] Flood R, McCartney M and Whitaker MAB 2008 *Kelvin: Life, Labours and Legacy* (Oxford: Oxford University Press) pp358
[30] Wilson DB (ed) 1990 *The Correspondence Between Sir George Gabriel Stokes and Sir William Thomson Baron Kelvin of Largs* (Cambridge: Cambridge University Press) pp783
[31] Smith C and Wise MN 1989 *Energy and Empire: A biographical study of Lord Kelvin* (Cambridge: Cambridge University Press) p5
[32] Thomson J 1825 (under the *nome de plume* Iota) Recollections of the Battle of Ballynahinch *The Belfast Magazine and Literary Jounal* **1** (Belfast: Morgan Jellett) p56-64
[33] Stewart ATQ 1977 *The Narrow Ground: Patterns of Ulster History* (London: Faber & Faber) p88
[34] Holmes RF 1981 *Henry Cooke* (Belfast: Chrsitian Journals Ltd) p13
[35] King ET (ed) 1909 *Lord Kelvin's Early Home* (London: Macmillan & Co Ltd) p1
[36] King ET (ed) 1909 *Lord Kelvin's Early Home* (London: Macmillan & Co Ltd) p61-62
[37] Stewart ATQ 1977 *The Narrow Ground: Patterns of Ulster History* (London: Faber & Faber) p98
[38] Barkley JM 1991 *Blackmouth & Dissenter* (Belfast: The White Row Press) p142
[39] Holmes RF 1981 *Henry Cooke* (Belfast: Chrsitian Journals Ltd) p54
[40] Holmes AR 2008 Presbyterians and science in the north of Ireland before 1874 *British Journal for the History of Science* doi:10.1017/S0007087408001234
[41] Thompson SP 1910 *The Life of William Thomson, Baron Kelvin of Largs* (London: Macmillan & Co Ltd) p32
[42] Dunlop J 1995 *A Precarious Belonging: Presbyterians and the conflict in Ireland* (Belfast: The Blackstaff Press) p28
[43] *Concise Dctionary of National Biography* 1961 (Oxford: Oxford University Press)
[44] King ET (ed) 1909 *Lord Kelvin's Early Home* (London: Macmillan & Co Ltd) p120-1
[45] King AG 1924 *Kelvin the Man* (London: Hodder and Stoughton) p123
[46] Thompson SP 1910 *The Life of William Thomson, Baron Kelvin of Largs* (London: Macmillan & Co Ltd) p626
[47] Thompson SP 1910 *The Life of William Thomson, Baron Kelvin of Largs* (London: Macmillan & Co Ltd) p627
[48] King AG 1924 *Kelvin the Man* (London: Hodder and Stoughton) p29-30
[49] Hutchison I 2009 *Journal of Physics (Conference Series)* – this volume
[50] Wilson DB (ed) 1990 *The Correspondance Between Sir George Gabriel Stokes and Sir William Thomson Baron Kelvin of Largs* (Cambridge: Cambridge University Press) p62-63
[51] Wilson DB 1974 Kelvin's scientific realisn: the theological context *Philosophical Journal Glasgow* **11** 41
[52] Nichol JP 1838 *The Phenomena and Order of the Solar System* (Edinburgh: William Tait) 64-65. Quoted in [51], p45.
[53] Paley W 1836 *Natural Theology* (2 volumes) (London: Charles Knight) pp386, pp391
[54] Thomson W 1894 *Popular Lectures and Addresses: Volume 2 Geology and General Physics* (London: Macmillan & Co) p204-205
[55] Wilson DB 1987 *Kelvin and Stokes: A Comparative Study in Victorian Physics* (Bristol: Adam Hilger) p79
[56] Thomson W 1894 *Popular Lectures and Addresses: Volume 2 Geology and General Physics* (London: Macmillan & Co) p199-200

[57] Thomson W 1894 *Popular Lectures and Addresses: Volume 2 Geology and General Physics* (London: Macmillan & Co) p198
[58] Thomson W 1894 *Popular Lectures and Addresses: Volume 2 Geology and General Physics* (London: Macmillan & Co) p204
[59] Thomson W 1894 *Popular Lectures and Addresses: Volume 2 Geology and General Physics* (London: Macmillan & Co) p464-465
[60] Thomson W 1894 *Popular Lectures and Addresses: Volume 2 Geology and General Physics* (London: Macmillan & Co) p45
[61] Wilson DB 1974 Kelvin's scientific realisn: the theological context *Philosophical Journal Glasgow* **11** p51

Journal of Physics: Conference Series **158** (2009) 012003 doi:10.1088/1742-6596/158/1/012003

Kelvin and industry in Ireland

Bernard Crossland[1] and Andrew Whitaker[2]
[1] Department of Mechanical Engineering, Queen's University, Belfast BT7 1NN, Northern Ireland
[2] Department of Physics, Queen's University, Belfast BT7 1NN, Northern Ireland

E-mails: b.crossland@qub.ac.uk , a.whitaker@qub.ac.uk

Abstract. Kelvin was a great mathematician, theoretical and experimental physicist, and educator, founding the first physical laboratory. He worked tirelessly for the creation of a reproducible set of physical units, and he was also an experienced and enthusiastic sailor. All these talents were linked to his extensive technological work, of which the most important examples were the laying of the Atlantic cable, and the marine compass. In Ireland his most important contributions were the occulting nature of the Holywood lighthouse, and his connection with the Giant's Causeway tramway. Kelvin's work on cabling and national maritime projects may have stimulated his later strong support of the British Empire and opposition to Home Rule in Ireland.

1. Kelvin – a man of many talents

Kelvin was certainly a man of very many talents [1-3]. From his Cambridge days he was a mathematician of the highest stature. Once he was established in Glasgow, he was able to broaden this talent, applying it to theoretical study of the physical universe, not only analysing a vast range of individual problems in the greatest depth, but also constructing major syntheses of tremendous range, and of at least potentially massive significance. Of these syntheses, thermodynamics has turned out to be the most important, while Kelvin's work on electromagnetism has also borne fruit through its development by Maxwell, and his work on models of the ether, while it must be judged less successful in the long term, served to raise many important questions and provide many useful techniques.

To this theoretical ability he swiftly added an exceptional experimental ability, both in the construction of novel and sophisticated equipment, and in the performance of difficult and important experiments. One of the most significant examples was the experimental checking in 1850 of the prediction made by his brother James (in the course of discussions with Kelvin himself) of the lowering of the freezing point of ice under pressure. This prediction and experimental confirmation are of the greatest possible importance, since they are based on the theory of Sadi Carnot, and are thus effectively the first results of the Laws of Thermodynamics, obtained, in fact, before these Laws were formally stated.

Kelvin described his experimental achievement to his Glasgow class as follows [4]: 'My brother James sent in a paper to the Society of Arts in Edinburgh stating that water freezes under a pressure of 760 mm at 0.0075 C below 0°. This has never been verified by any experiment till I did it the other day by myself, with a most delicate thermometer. This thermometer is assuredly the most delicate that ever was made, there being 71 divisions in a single degree F. It is filled with ether.... mercury would not do. .....Under a pressure of 8.1 atmospheres, the temperature was lowered $7\frac{1}{2}$ divisions. 0.106 is the lowering by experiment: 0.109 by theory. This is within

1/3000 of a Fahrenheit degree.' This triumph came little more than 3 years after he took up his chair, at which time he was practically a novice from the experimental point of view.

In the educational field, while his own lecturing was probably more successful in stimulating the very best students than in communicating effectively with the average ones, he has his deserved fame as the founder of the first student physical laboratory, rather different from the usual style of today in that students voluntarily carried out practical investigations on topics of interest to Kelvin, funded by small grants obtained from the Royal Society. With this experience, Kelvin was able and willing to advise on the new physical laboratories at Oxford and Cambridge, the Clarendon and Cavendish respectively.

Kelvin also played an extremely important role in the setting up of a system of electrical and other units; this was clearly an absolutely essential condition for the production of accurate and reproducible scientific results, for the statement of technical requirements in industry, and for any form of marketing of electricity. Work on units demanded first a detailed understanding of how accurate and reproducible standards might be set up, the various difficulties and errors that are certain to occur and how they may be overcome. It also demanded effective committee work, taking the views of scientists and engineers from different nations, and tactfully but firmly arguing for the most effective conclusions. Indeed Kelvin was an excellent member and chair of important committees, particularly a number of committees to do with naval matters, establishing what had gone wrong when ships had been lost at sea, and helping to plan future generations of naval vessels.

This takes us to another of Kelvin's talents, one that seems, at least at first, more a matter of relaxation than those described so far, sailing. By 1870 he was well enough off to be able to buy a substantial schooner-yacht, the *Lalla Rookh*, which was large enough in fact to require a professional master and crew. From then on, in the yearly six-month break from duties in Glasgow, Kelvin would undertake cruises, both short, around the coast of England and Scotland or to visit family and friends in Belfast, or very much longer, such as the trips to Lisbon, Gibraltar and Madeira. The yacht was a residence, providing him with sufficient peace for work, a laboratory, where he was able to interact directly with the waves that he studied theoretically in the winter months, and an engaging and adventurous hobby. It taught him a great deal, of course, about maritime affairs.

2. Kelvin and the practical application of science

Each of these talents may be described as a piece in the jigsaw of Kelvin's life and work, but they were certainly not independent, and we have not mentioned as yet the jigsaw piece that connected them all together. This was Kelvin's work as a technological inventor and entrepreneur. It was a facet of his life that began in the mid-1850s, and came to occupy an increasing fraction of his time, bringing him fame and fortune. The fame included his knighthood. (For his peerage, see the article *Lord Kelvin and liberal unionism* by Iain Hutchison in this collection.) The fortune amounted to at least many hundreds of thousands of pounds, perhaps in today's terms many tens of millions of pounds. This in turn enabled Kelvin to own Netherhall, his substantial mansion in Largs, and, of course, the *Lalla Rookh.* It also enabled him to show substantial generosity to Glasgow University, to Peterhouse, his college in Cambridge, and almost certainly to many other people and institutions.

It is important to stress how much Kelvin's technology relied on all aspects of his physics and mathematics. Let us first consider his work on telegraphy. It is, of course, well-known that it was Kelvin's work on the Atlantic telegraph that finally brought success to this venture in 1866 after many costly failures. His work in telegraphy [5] actually began from purely academic discussions with Stokes on some ideas of Faraday, and much of the reason for Kelvin's involvement with cable-laying was the fact that it seemed by no means obvious to many of those involved that such considerations were at all relevant.

Kelvin's main protagonist, Wildman Whitehouse, was a medical man not a scientist or engineer, but nevertheless he was designated "electrician" on the 1857-8 attempts to lay an Atlantic cable, and as such he was responsible for all electrical matters, until his summary dismissal after the failure of the 1858 cable. It was obvious that he was an ignoramus; he clearly and categorically separated theoretical matters, with their place in the lecture room, from the practical experience accumulated, he felt, by people like himself. What is more striking, though, as pointed out by Crosbie Smith and Norton Wise [6] is that even such undoubtedly excellent electrical engineers as Lewis Gordon, Professor of Engineering at Glasgow, and indeed the first such professor in Britain, and Werner Siemens, founder of the Siemens empire, "did not approach their work from a thoroughly articulated theoretical perspective".

It was Kelvin who stressed the importance of physical argument in the design of telegraph cables, and this focused attention on such parameters as the resistance and capacitance of the cable wire. In mid nineteenth-century, however, such attention tended to be abortive because of the absence of reproducible standards for the basic electrical units. The lack of such standards meant that it was impossible for those planning cabling expeditions to demand cable of determined standard from manufacturers. And, of course, it was this requirement that led to Kelvin's extreme interest for the remainder of his life in the establishment of a set of standards [7, 8].

While Kelvin thus stressed the importance of theoretical ideas for technical purposes, in particular for cabling, he made another massive and completely different contribution to this area. This was his ability to construct instruments that could not only work effectively, but would do so in the most difficult of circumstances. A very important example is Kelvin's mirror galvanometer. Its sophistication briefly saved the day for the 1858 expedition when Whitehouse surreptitiously substituted it for his own crude instrument. The Board remarked that Whitehouse's investigations had cost the company £12000, but would have rendered it a laughing-stock had not the Board been "fortunate enough to have an illustrious colleague who had devoted his mind to the subject, and whose inventions produced *in his own study – at small expense* and from his own resources are available to supersede the useless portions of apparatus prepared at great labour and enormous cost for this special occasion" [9].

Equally significant was Kelvin's siphon recorder, the first instrument to provide automatic recording of telegraph signals on moving tape, thus saving immense amount of operative time watching and recording the fluctuations of galvanometers. Even more interesting, perhaps, was that Kelvin also worked on a completely different aspect of cabling, that of laying the cable in such a way to avoid it breaking, a very difficult task, of course, when seas may be fierce. Kelvin's invention, based on the technique of fly-fishing, had the potential to save large amounts of time, effort and money for the cabling company, and all these inventions were also eventually, of course, to prove exceptionally lucrative for Kelvin and his associates.

Exactly the same combination of inputs was involved in what was probably Kelvin's second most important and lucrative set of inventions, those associated with the marine compass [10]. Kelvin's work started with a highly theoretical analysis of the magnetism of ships – essentially to study the joint effect of the magnetism of the earth and that of the iron in the ship. This work was developed from earlier studies of Poisson, and also Kelvin's friend, Archibald Smith. It also involved incisive study of the mechanics of the various components of the compass. The final stage of this work was the far more pragmatic matter of ensuring that the compass would work effectively, giving a quick and accurate reading even in the most atrocious weather conditions.

It is clear that all aspects of Kelvin's library and laboratory based "pure research" impacted directly on his technical and entrepreneurial work, but, of course, this was a two-way process. Ideas and problems generated while carrying out practical applications constantly fed back to innovative investigations in the laboratory.

Kelvin's developments in teaching also related directly to his technical work. In fact it was precisely the requirement to provide the necessary data to enable industry to use physical theory effectively that led him to encourage undergraduate students to carry out highly relevant research.

While this appears a somewhat pragmatic approach, it is certain that Kelvin came to value greatly the teaching of science and engineering via the science laboratory, which was so important in developing a physical understanding of the science and a feel for its practical applicability.

Kelvin's protégé in this approach was John Perry. Perry was an extremely important and interesting character, who may best be described as a science-based engineer, who interacted with Kelvin at several periods of his life. He makes a number of appearances in the article by Andrew Whitaker, *Is Kelvin representative of an "Irish tradition"?*, in this volume, and there is a longer account in an collection to be published soon [11]. For the present purpose, Perry's importance is that he was exceptionally keen on the practical teaching of physics and engineering. As mathematics master and physics lecturer at Clifton College in Bristol from 1870-84, he was to set up the first physics laboratory and the second mechanics workshop in an English school. Later, in Japan and back in London, he was to demonstrate the same commitment.

Lastly we turn to Kelvin's devotion to maritime affairs. What may have seemed initially just a pleasant hobby can now be seen to be an essential part of his approach to life. His voyages not only showed him clearly the perils and difficulties of life at sea, but also gave him magnificent opportunities for testing possible solutions. Probably the most important example of this is his apparatus for sounding, or ascertaining the depth of water while at sea, clearly an essential task in order to avoid running aground with great danger of loss of life. The considerable advantage of the apparatus produced by Kelvin was not only that the procedure may be carried out efficiently and accurately, but that the ship need not be halted to carry out the sounding, thus avoiding the loss of time, or indeed the temptation to sail or steam on to avoid such loss of time.

Very many of Kelvin's inventions were connected with the sea, and it should not be thought that his only motivation was to make money. As a sailing fanatic himself, he certainly and genuinely wished to reduce loss of life, remarkably common through at the very least the first 65 years or so of the nineteenth century. Silvanus Thompson [12] quotes "the highest authority in the British navy" as saying that he considered Kelvin to be "the man who had done by far the most for the advancement of navigation" in that period, and he also quotes a sailor as saying that "I don't know who this Thomson may be, but every sailor ought to pray for him every night".

Kelvin produced a very wide range of inventions; he wrote, in fact, 70 successful patents [13]. He came up with the idea of the heat pump which was, in principle at least, an extremely cost effective way of causing a transfer of heat from the cold exterior of a building to the warmer interior. (A heat pump is essentially a refrigerator acting in reverse.) More prosaically he invented a very well-known non-drip tap. Rather than further general description of this work, for the remainder of this paper we concentrate on that work closely connected with Ireland, and then finally mention the effect that the success of much of his technical work may have had on his political views.

3. The Holywood Bank lighthouse in Belfast Lough

In May 1873, Kelvin wrote to Thomas Andrews, an old friend and Vice-President and Professor of Chemistry at Queen's College Belfast, replying to an invitation to stay with him and his wife. Kelvin hoped to be able to stay for a night when he travelled over to Belfast to set up the eclipsing arrangement that had been ordered by the Harbour Commissioners for the lighthouse on the Holywood Bank [14]. By April 1875 he was able to report [15] that: "I am told by the harbour authorities that captains, pilots, and sailors are all much pleased with [the lighthouse]. They know it with perfect confidence from any other light afloat or ashore."

Kelvin's interest in lighthouses was first demonstrated in an article he wrote for *Good Words* in 1873 [16], while many of the ideas were amplified in a lecture given at the Naval and Marine Exhibition at Glasgow in 1881 [17]. Lighthouses were naturally a topic of great interest to him, since they combined technical ideas with safety considerations at sea. For Kelvin, a lighthouse must fulfill two functions: first it must be seen, and secondly it must be recognised, by which one means distinguished from any other light in a sufficiently large region of sea. He felt that the system of lights in 1873 failed mariners in many ways, and indeed that there was little improvement by 1881.

In both years the very great majority of lights were either *fixed* or *flashing*. In 1873 the count was 490 out of 623 fixed and 112 flashing. Fixed lights have the obvious positive point of always being visible, thus fulfilling the first function, but, as long as they transmit white light, equally obviously fail to fulfill the second function, as all lighthouses will appear identical. In a very few cases red light was used, but Kelvin considered this a highly retrograde step. Red lights were easily confused with the red side lights of sailing ships, and also were not easily visible in fog. (Kelvin did admit that, when used in conjunction with a white light, red light could be useful for indicating a particular direction.)

When categorising lighthouses, a flashing light is taken to be one where the period of light is much less than that of darkness. In fact in nearly all cases the *flashing* is caused by the light revolving, so that an observer in any particular direction will see the light appearing to flash. (In fact Kelvin uses different names in his two papers; in the 1873 paper, what we are following his 1881 usage in calling *flashing,* was called *revolving*, while the term *flashing* was used for a very small number of rather different sets of light we will mention shortly.)

While such lights could be cost-effective, in the sense that the light was produced in a narrow but intense beam, Kelvin considered them highly ineffective from the point of safety. The problem was compounded by the fact that in nearly all cases, the overall period was long, of the order of minutes, and the period of light was a very small fraction of that period. Kelvin wrote of the sailor in awful weather or very low visibility who gets a glimpse of a light, and then has to peer into the mist hoping to catch another glimpse some minutes later.

One possible perceived merit of the flashing (revolving) light, one "defence" of the existing system as Kelvin calls it, is that the period of the light may be used to characterise the particular lighthouse. "This defence", Kelvin says, "is utterly invalid. It is scarcely possible for any one, counting time to himself on a gusty, showery night at sea, to distinguish a one-and-a-half minute from a two-minute revolving light." One general recommendation of Kelvin was that all periods of any time sequence of light should be short, so that the sailor does not have to wait searching for the reappearance of the light. In the 1873 paper the term *flashing* was reserved for the rare lighthouses where there are five or more flashes within a minute, thus at least satisfying this criterion.

A nice quotation that Kelvin gives to this effect is from Captain Moriarty, the man who, in mid-ocean in 1866, was able to find the end of the Atlantic cable that had been lost the previous year, and was thus undoubtedly, as Kelvin says, "a very practical man". While making for the Irish coast in dirty weather, Moriarty said that "Those lighthouses should flash out their characters like your electric signals; every lighthouse should flash, and flash and flash, many times in a minute, showing you which lighthouse it is every time. That long minute of the revolving light has often seemed to me like an age, when I have been anxious to make out where we were in a gale of wind and rain."

Even flashing lights with a short period, however, still have two defects: the dark period is longer than the light period, and there is still the question of distinguishing different lighthouses. Here Kelvin's suggestion was a development of the idea of the famous mathematician Charles Babbage made as early as 1851, though Babbage's note had been pigeon-holed by the Admiralty. This was the method of the *eclipsing* or *occulting* light. In this method, the period of the light is much greater than that of the dark. This certainly makes it easier for the sailor to catch sight of the light in difficult conditions. Kelvin believed that the presence of a short dark period actually made the light easier to see than a fixed light, because the switching on catches the attention of the watching sailor.

For recognition of lighthouses, Babbage and Kelvin suggested breaking the eclipse itself into characteristic sets of short periods separated by periods of light. Babbage's suggestion was unwieldy – for lighthouse number 347, for example, the periods of eclipse would be in groups of

3, 4 and 7. Kelvin's approach was much neater: there would be a group of eclipses consisting of short and long dark periods. There might be a single long eclipse of 3s in an overall period of 12s (i.e. 75% light). There might be a single short eclipse of $\frac{1}{2}$ s in an overall period of 10s. In all other cases a short eclipse would be 1s and a long eclipse 3s, and they may be combined to give, for example, short-short or short-long or long-short-short; in 1881 Kelvin lists ten possibilities in all. In each case except the single short, the time from beginning to end of the group of ellipses would be about $\frac{1}{3}$ of the total period.

It is clear that Kelvin's proposal was based on the Morse code. For example the Holywood Bank lighthouse would be distinguished as short-short-long or dot-dot-dash, which is just U in Morse code. By 1881, Kelvin thought it unwise, though, to stress this aspect of his plan, as it might frighten "practical" men! Actually in 1873 Kelvin had listed all 26 codes (figure 1)

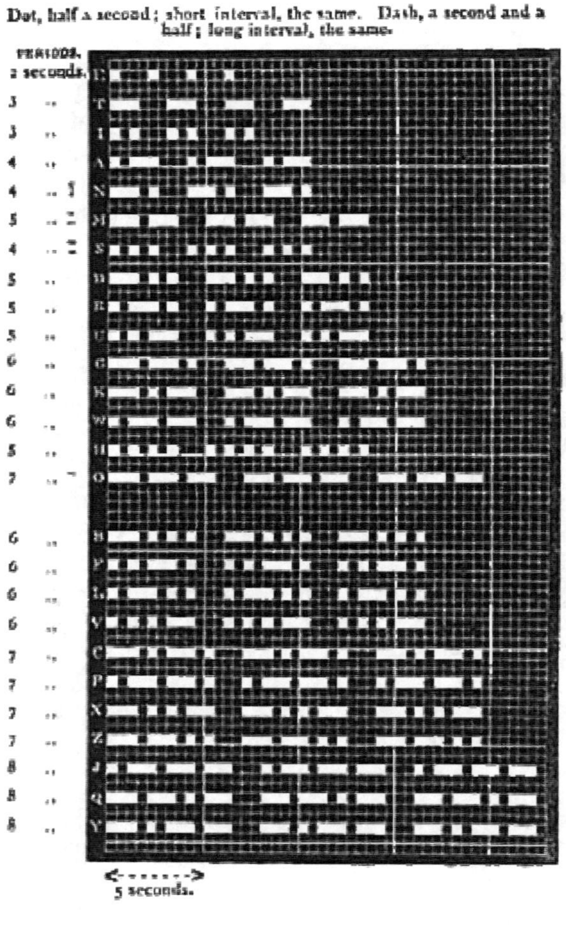

Figure 1
Kelvin's full list of groups of light and dark periods presented with the corresponding Morse codes in 1873. For example the tenth down is U: short-short-long, or dot-dot-dash, the sign for the lighthouse on the Holywood Banks.
By 1881, Kelvin had removed the stress on Morse codes as liable to frighten "practical men" and he had decided to concentrate on the ten simplest patterns

Kelvin added that the same idea might be applied to the fog-siren also used at lighthouses. In this case, though, the distinction between short and long would not be satisfactory, as echoing may make the distinction difficult to appreciate. However this distinction may be replaced by one between a high note and a low note, which, he reports, is readily appreciable to the least musical ear.

Kelvin discussed the practical means of achieving the eclipsing of light, using oil or gas; gas was quite convenient and economical, because it could just be turned down for the dark periods.

He recognised, though, that the future for lighthouses must lie with electricity. He wrote that: "The great adaptability of the electric light to furnish increase of power when wanted gave it a value which no other source of light possessed" [18]. In his 1881 paper he reported that he had developed a simple and inexpensive apparatus that was able to produce any of the groups of eclipses. His machine was applicable to any lighthouse, large or small, and, by appropriate optical arrangements, it could channel the light to the horizontal plane or in a particular direction as required. This machine, he said, had been at work for a month in the college tower of the University of Glasgow, performing for four hours an evening without any sign of wear.

Kelvin began his 1873 paper by presenting a number of accounts of cases, imaginary and real, where the failures of the present system of lights could lead or had led to disaster; one of the real cases had led to 183 deaths. By 1881 there had been little enough improvement. Kelvin considered that the responsible authorities were too little concerned with the issue to take on the necessary work and expense. Sailors, of course, looked on lighthouses as works of nature rather than art, and "would as soon think of asking to have the shape of Knockdolian changed, so that it could never be mistaken for Ailsa Craig" as to request the changes suggested by Kelvin, which might actually save their lives.

However Kelvin had been given his head with the Holywood Bank lighthouse at the head of Belfast Lough [15]. Until November 1874 it was a red fixed light that could only be seen from two miles off and was constantly liable to be mistaken for a ship's port-side light. At that time the red shade was removed and Kelvin's eclipsing machine was attached. Besides having the advantage of being unmistakeable for anything else, it was then visible for distances up to five miles away.

Subsequently Kelvin reported the following events [17]. In the small hours of a summer morning, he was approaching Belfast Lough with a celebrated lighthouse engineer. They saw a light, which Kelvin asked his companion to identify. His companion could not do so, and for the first time admitted that it was possible to confuse a lighthouse light with one on a steamer's masthead. It was, in fact, as they subsequently learned, the Copeland lighthouse off the south entrance of Belfast Lough, which is a fixed light. Soon afterwards they saw the Holywood Bank lighthouse, which was actually just visible a full ten miles away. With its distinctive short-short-long pattern of eclipsing, its identity was immediately clear, together with that of the until then unidentified fixed light they had seen before. Kelvin's theoretical ideas had, as he always wished, been proved to work in practice.

4. The Giant's Causeway tramway.

Kelvin became interested in the use of electricity for light and power in the late 1870s [19]. He argued that waste coal burned at the pit head, or hydroelectric energy at Niagara Falls could be used for the creation of electricity using dynamos, and this could then be transmitted hundreds of miles. Not only the economy but also public health would benefit.

In 1890 he chaired the Niagara Commission to discuss the way in which the power of Niagara could be utilised for the production and transmission of electricity. While many of the recommendations of the Commission were taken up, at least in part, the part of Kelvin's position that has become best known is his steadfast opposition to alternating current; he felt direct current should be used. He did admit that, for the sake of economy, extremely high voltages were required for electric transmission over long distances, and, of course, it is that feature which is usually said to make the use of alternating current essential, since with alternating current transformers can be used to transform the voltage to the low value required for safety in the home or factory environment. Kelvin's argument for using direct current was that, when using alternating current, the average size of the voltage is reduced by a factor of $\sqrt{2}$ over the direct current value with the same circuit components.

Journal of Physics: Conference Series **158** (2009) 012003 doi:10.1088/1742-6596/158/1/012003

Kelvin became involved with the company set up by Joseph Swann to manufacture the Swann glow-lamp. So excited was he by the idea of electric light that in 1881, he arranged to have his University residence, his laboratory and his yacht lit by Swann lamps, a Clerk gas-engine and Faure cells. He was particularly interested in the use of the Faure cell for the storage of electrical energy, and expended much research time on this project. He suggested that a source of electrical energy such as a windmill might be made practicable by the use of Faure cells to store some of the energy which could then be used in calm periods. Well over a century and a quarter later, the problem of efficient storage of energy is still unsolved; windmills are very much with us, but there are still potential major problems in feeding their energy into the electric grid.

Also in the 1880s, Kelvin became associated with the budding engineer and entrepreneur, Sebastian de Ferranti. Actually the cause of their involvement was the fact that a patent for an alternator that Ferranti obtained in 1882 was practically identical with a machine Kelvin had designed in the previous year. Kelvin remained associated, primarily financially, with Ferranti for a long period, as the young man (born only in 1864!) survived early calamities, to move on to become a major provider of electric light by the end of the 1880s, and of course his name is still famous in today's industry, particularly in the area of defence.

With this great interest in the use of electricity for light and power, it is not surprising that Kelvin took the opportunity to become involved in an interesting project in his "beloved Ireland". This was the Giant's Causeway tramway [20, 21], the first electric tramway in the United Kingdom, and the first in the world to be powered by hydroelectricity.

The progenitor of the tramway was William Acheson Traill (1844-1933) [22], who was born in Bushmills, around two miles from the famous tourist attraction, the Giant's Causeway, and educated as a civil engineer at Trinity College Dublin. Traill was well-acquainted with Sir William Siemens, the member of the Siemens dynasty who had travelled to and worked in England. At the 1879 Berlin Exhibition, the Siemens brothers had demonstrated a working electric railway using their own dynamo and motor, and when Traill and his brother Dr Anthony Traill proposed a tramway to link the Causeway via Bushmills with the popular seaside resort of Portrush, Siemens Brothers Ltd took £3500 of shares, and Sir William Siemens was given a seat on the Board as technical adviser. Kelvin also took out shares of value £1000; this might be worth around £100,000 today, so he must be regarded as a major shareholder.

Construction began in 1881, with Siemens providing much of the expertise and equipment, and operations started in 1883, though initially only from Portrush to Bushmills. At first sight this seems strange, as one would have assumed that the main purpose of the tramway was so bring people to the Causeway; however from the start a connection from Bushmills to the Causeway by horse car was available. The main source of power was hydroelectricity generation at Walkmills on the River Bush near Bushmills, though this was augmented by a steam-powered generator at Portrush. In the usually damp and salty atmosphere of North Antrim, it was discovered during construction of the tramway that two rails could not provide the necessary current, so a third higher conductor rail was required. This third rail was too dangerous to be used in the towns of Portrush and Bushmills so a steam locomotive was available to work these parts of the line, and also to convey goods and minerals. The system was designed to work at 225/250 volts direct current. Rolling stock consisted of first and third class coaches, some open and some closed, and also goods and mineral wagons (figure 2).

The official opening was not held until September 1883. At this ceremony it appears that Kelvin was in exuberant mood, shaking hands with various of the guests while holding a live wire in his other hand. He apparently believed that electric shocks were a cure for many illnesses, and local doctors were prepared to back up this claim for such conditions as rheumatism. Several members of the Board also dropped their trousers and sat on the live conductor rail in order to demonstrate the safety of the system to the public and, most importantly, to the Inspectors. (Years later Traill's daughter asked him: "Did sitting on the rail not hurt?" He replied: "It hurt like blazes, but we weren't going to let the Inspectors know that!")

Clearly Kelvin was already taking practically a proprietorial interest in the tramway as a result of his large investment and also his very close relationship with Sir William Siemens. Sadly only two months later, Siemens suddenly died. Kelvin was devastated; Lady Thomson wrote to a friend: "We have been at Sir W Siemens's funeral today. His death has been a great grief to [Kelvin] – so much so that I felt I must come up with him and look after him" [23]. Kelvin replaced him on the Board and also as technical adviser.

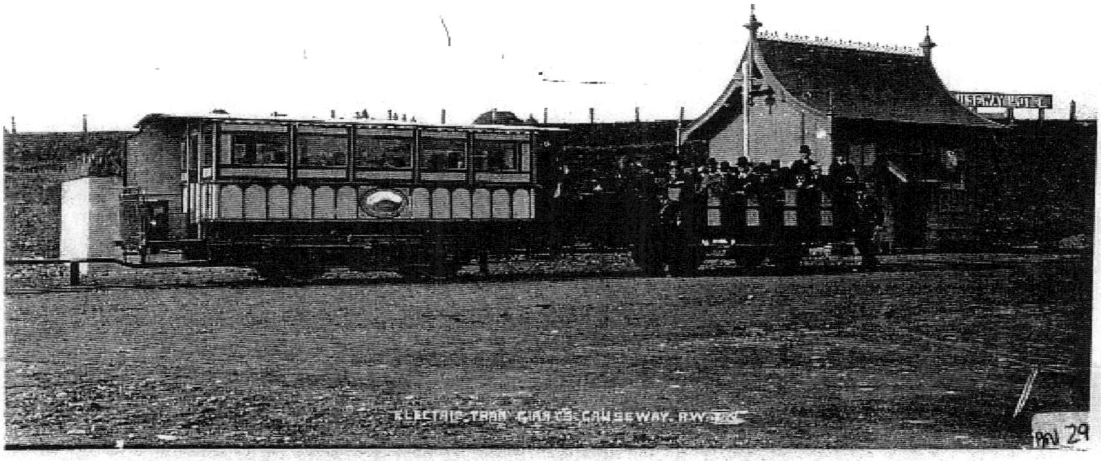

Figure 2.
The Giant's Causeway tramway in operation.
[The Ulster Museum: Welch Collection]

The good news was that the tramway was a commercial success. At the opening Traill was able to report that, in the 12 weeks since the first of July alone, over £1000 had been taken, and by 1887, the Board felt able to increase the length of the tramway by a further mile and a half, taking it on to the Giant's Causeway as originally planned. This required the building of a relatively substantial bridge across the River Bush, and it may have been this requirement that cause the initial limitation to Bushmills.

A trip from Portrush to the Causeway and back, with the section from Bushmills initially covered by horse car, cost 1/9, $8\frac{3}{4}$p in today's terms, though the equivalent to 1/9 today might be more like £10. It seems clear that the main clientele were holiday-makers and trippers, prepared to pay a good sum for a dramatic trip through sometimes breathtaking scenery. Throughout the life of the tramway, the summer months were always by far the busiest, and indeed from the mid-1920s the tramway shut down in the winter. Most of the employees were laid off for the winter, but fortunately these months were the prime ones for employment in Bushmills Distillery.

Traill was frequently to write to Kelvin with requests for advice [24]. Not surprisingly, with a new technology, there were many teething problems with the electrical systems, and there were additional constraints due to the situation of the electric rails very much adjacent to people and animals. Despite the death of Siemens, Kelvin was able to negotiate maintenance of the favourable terms with the Siemens Co. for technical improvement and repair. When the Bill for the extension was sought, there was an Inquiry and it was Kelvin who took the brunt of the technical questioning, particularly with regard to the safety of the hazardous third rail.

For such a small undertaking, the Giant's Causeway Tramway actually had quite a long period of useful life, but eventually it was to close in September 1949. However it is pleasant to report that the mile and a half stretch from Bushmills to the Causeway was reopened in Spring 2002, the operation being run by a not-for-profit organisation (figure 3). Incidentally, to compare with the prices above, a return on this line today costs £3.50.

Figure 3.
The Giant's Causeway line in use today

5. Kelvin: engineering and imperialism

In 1798, the United Irishmen took up arms against the government [25]. The aim was to unite 'Catholic and dissenter' – effectively Catholic and Presbyterian - to demand complete religious equality and radical reform of Parliament. In broad terms the opposition in Ireland was the established Irish church, which was actually the Irish branch of the Church of England, many of whose members might be categorised as Anglo-Irish land-owners. The rebellion was crushed without mercy. James Thomson, Kelvin's father, was witness to the Battle of Ballynahinch, in which the rebels were routed; after the battle the Crown forces took the opportunity to burn, wreck and loot the town of Ballynahinch. In the words of Smith and Wise [25]: "James Thomson's [later] account of the battle emphasised the futility of the rebellion, while at the same time expressing an implicit sympathy with the ideals of the rebels – the ideals of liberty, equality, and freedom from sectarian dogmatism – and an explicit hatred of the repression and atrocities perpetrated by the establishment forces." Doubtless these were views he passed on to his children.

Yet, as Kelvin's career developed, while he retained his religious toleration, he inevitably became closer and closer to what might be called the "establishment" of the British Empire. His achievement with the first Atlantic cable, of course, made him famous and brought him his knighthood, but this was only the beginning of his successes in cabling. Bruce Hunt [26] has written that: "Submarine telegraphy was one of the characteristic technologies of the British Empire in the second half of the 19th century, and well into the 20th, and William Thomson sat right in the middle of it. The most famous cables were those across the North Atlantic, but there was an extensive network running throughout the rest of the globe, particularly linking the parts of the British Empire. Almost all of these cables were built, laid and operated by British companies, and between the 1870s and 1900 *all* of them used instruments designed and often manufactured by Thomson and his partners."

Equally significant was his work on navigation and sounding, in particular his magnetic compass [10], which was first used universally on merchant vessels, and then became the standard compass for use in the Royal Navy. The size of the market may be judged by the fact that, in the decade after Kelvin's death, over 3000 compasses of his design were sold. It was no idle remark that in the second half of the nineteenth century, the British navy ruled the waves, and Kelvin may justifiably have claimed that his maritime inventions had a large part in achieving this. During his campaign of persuading the Admiralty to adopt his compass, Kelvin naturally became on close terms with many of the most important men in the Navy.

Almost inevitably Kelvin's contributions to the Empire through cabling and through his maritime inventions led to his becoming a strong supporter of the Empire. Thus the scene was set for one of the last adventures of his life. As an Irishman, he believed that the future for Ireland should be as a full participant in the British Empire. One of his arguments brought together politics and engineering; via the telegraph Dublin and London could communicate "at the rate of 500 words per minute" and hence a separate parliament was not needed [27]. Thus, though a Liberal to that point, he broke with Gladstone over Home Rule. This story is told by Iain Hutchison in this volume.

References

[1] Thompson S P 1910, *The Life of William Thomson: Baron Kelvin of Largs* (London: MacMillan) (2 vols)
[2] Smith C and Wise M N 1989 *Energy and Empire: A Biographical Study of Lord Kelvin* (Cambridge: Cambridge University Press)
[3] Flood R, McCartney M and Whitaker A, ed 2008 *Kelvin: Life, Labours and Legacy* (Oxford: Oxford University Press); chapter 6, Kelvin and engineering by B Crossland, is especially relevant for the present paper.
[4] Ref. 2, p. 282. They are taken from William Smith, Notes of the Glasgow College natural philosophy class during the 1849-50 session.
[5] Ref. 2, chapter 13, Telegraph signals and light waves,: Thomson versus Maxwell, pp 446-63; chapter 19, The telegraphic art; and also chapter 20, Measurement and marketing, pp 698-712
[6] Ref 2, p 666
[7] Tunbridge P 1992 *Lord Kelvin: His Influence on Electrical Measurements and Units* (Stevenage: Peter Peregrinus)
[8] Ref 2, pp 684-98
[9] Ref 1, p 370
[10] Ref 2, chapter 22, The magnetic compass
[11] Crossland B 2008 John Perry *The Lives of Great Engineers of Ulster* vol 3 ed B Crossland and J S Moore (Belfast: NE Consultancy and The Bookshop at Queen's)
[12] Ref 1, p 735
[13] Ref 1, pp 1275-7
[14] Ref 1, pp 635-6
[15] Ref 1, p 659
[16] Thomson W 1873 (March) Lighthouses of the future *Good Words* 217-24; also in http://zapatopi.net/kelvin/papers/lighthouses_of_the_future.html

[17] Thomson W 1891 On lighthouse characteristics *Popular Lectures*: *Vol 3 Navigational Affairs* (London: Nature) pp 389-421; also in http://zapatopi.net/kelvin/papers/on_lighthouse_characteristics.html
[18] Ref 1, p 713
[19] Ref 2, chapter 20, Measurement and marketing, pp 712-22; this is a good reference for the whole first part of this section
[20] McGuigan J 1964 *The Giant's Causeway Tramway* (Lingfield: Oakwood)
[21] McGuigan J 1983 *The Giant's Causeway, Portrush and Bush Valley Railway and Tramway Co. Limited* (Holywood: Ulster Folk and Transport Museum)
[22] Crossland B and Moore J S 2003 *The Lives of Great Engineers of Ulster* vol 1 (Belfast: NE Consultancy for Belfast Industrial Heritage) chapter 23, pp 166-72
[23] Ref 1, p. 799
[24] Ref 2, p 720
[25] Ref 2 pp 7-8
[26] Hunt B 2008 (January) Kelvin the telegrapher *Institute of Physics History of Physics Group Newsletter* **23** 7-16
[27] Ref 1 pp 869-70

Journal of Physics: Conference Series **158** (2009) 012004 doi:10.1088/1742-6596/158/1/012004

Lord Kelvin and Liberal Unionism

Iain Hutchison
Department of History, University of Stirling, Stirling FK9 4LA
Scotland

E-mail: iain.hutchison@btinternet.com

Abstract. Kelvin had been a staunch Liberal until Gladstone's Irish Home Rule Bill of 1886 made him a Liberal Unionist. He believed the measure would damage Ireland's best interests and would create economic and social instability. He also rejected it on constitutional grounds. Concern for the position of Ulster's Protestants was, however, not a major issue for Kelvin. Between 1886 and 1890 he was the most prominent Liberal Unionist in the West of Scotland, and under his leadership, the party was highly effective in the region. It is likely that, to a certain extent, his peerage was a reward for his political work. After 1890, Kelvin became more of a party figurehead, while his views on many political issues drew him closer to those of the Conservatives.

1. The background to the 1886 Irish Home Rule Bill crisis

On 8 April 1886, the Liberal Prime Minister, William Ewart Gladstone (figures 1 and 2), revealed his proposed legislation to grant Home Rule to Ireland. The main elements of this legislative package were that an Irish parliament would be established in Dublin, and as a consequence no Irish constituencies would be represented in the Westminster Parliament. The Dublin parliament would have responsibility for domestic legislation relating to Ireland, while Westminster retained control of foreign and Imperial policy, along with the armed services, and certain fiscal matters. Additionally, an ambitious scheme to buy out Irish landowners and transfer ownership to tenant-farmers was brought forward as an integral part of the scheme.

Gladstone had a number of motives in proposing Home Rule for Ireland, and, as was quite common with him, they were a blend of high-minded principle and low political cunning [1]. The problems of Ireland had preoccupied British statesmen at regular intervals since 1800; Gladstone calculated in 1886 that for 42 of his 54 years in politics, Irish matters had been at the forefront. As he wryly remarked to Alfred, Lord Tennyson about his political career, 'Ireland has had a rather dominant influence over it'. For nearly twenty years before 1886, Gladstone had been actively engaged in seeking to definitively settle the question. He famously announced when forming his first ministry after winning the 1868 general election, "My mission is to pacify Ireland." Yet since then the urgency of the Irish crisis had intensified, not diminished, despite his various legislative initiatives over the next two decades. His second administration, lasting from 1880 to 1885, had been bedevilled by the political demands of Irish nationalism, the problems of Irish agrarian discontent, and the obstruction of parliamentary business by Parnellite MPs (figures 3 and 4). As he ruefully remarked, this ministry had found Irish issues "especially (and rather unexpectedly) prominent."

Journal of Physics: Conference Series **158** (2009) 012004 doi:10.1088/1742-6596/158/1/012004

Figure 1. William Ewart Gladstone (1809-98), British Prime Minister from 1868-74, 1880-85, Feb to July 1886 and 1892-94. From 1885, he was convinced that Home Rule was the solution to the Irish question.

Figure 2. William Gladstone in the 1880s, as drawn by the artist Phil May.

Until the early 1880s he had been adamant that there could be no constitutional dilution of the full union of 1800 between Britain and Ireland: self-government was not for him a political option. This position he had stated explicitly and fully in various public forums on many occasions from the mid-1860s. His mind, however, had been moving to Home Rule from some point in 1882, and by the first half of 1885 he seems to have definitely decided that this was the only solution. He concluded, after thorough study, that the union had been harmful to Irish development and stability, so drastic remedies

Figure 3. Charles Stewart Parnell (1846-91), founder and leader of the Irish Parliamentary Party and, as such, 'uncrowned King of Ireland'. A crisis developed in 1890 when his relationship with Mrs Katherine O'Shea became known and the party split. In September 1891 he gave a crucial speech in pouring rain, but caught pneumonia and died.

Figure 4. The Irish Parliamentary Party in April 1886. Parnell is standing in front at left of centre.

Figure 5. Joseph Chamberlain (1836-1914), leader of the Radical wing of the Liberal party, and contender for the leadership. He opposed Gladstone over Home Rule. As a Liberal Unionist, he was to play an important part in the Conservative governments of Lord Salisbury and Arthur Balfour.

Figure 6. The Marquess of Hartington (1833-1908), leader of the Whigs or moderate Liberals; he also opposed Gladstone. He became Duke of Devonshire on the death of his father in 1891, and was to serve in the Conservative administrations of Lord Salisbury and Arthur Balfour.

were now inevitable. Nothing but a broad, generous measure of self-determination would yield a lasting solution. The general election of 1885 added significantly to Gladstone's resolve to introduce Home Rule. Parnell's party increased its numbers at Westminster from 59 to 86 MPs, wiping out all Irish Liberal MPs. The nationalists now represented fully five-sixths (85 [2] out of 103) of Irish constituencies – a democratically achieved majority which could not easily be ignored by politicians in the rest of Britain. Gladstone claimed that after this verdict at the ballot box, there was a very real danger that Irish nationalists might turn to violence if their demands were not met. As Ireland had been racked for nearly a decade by agrarian unrest, and Fenian "outrages" were a contemporary reality, this was a compelling argument for many.

Figure 7. Gladstone moving the second reading of the Home Rule Bill in the house of Commons on May 10th 1886. Parnell is in the second row on the right.

But there were other factors behind his decision. Gladstone, a vain man, regarded the solution of the Irish question as the last and greatest task in his distinguished political career, and he believed that he alone was capable of tackling it successfully. It has also been argued that he feared that the Liberal party would split if he was succeeded by the Radical contender for the leadership, Joseph Chamberlain (figure 5). Chamberlain's advocacy of radical policies, such as the disestablishment of state churches, was viewed with extreme hostility by the moderate, or Whig, wing of the party, and a large number threatened to secede if the Chamberlain programme were introduced – including, as shown below, many in Scotland. One successful device adopted over the years by Gladstone to maintain Liberal unity was to get the party to focus on one over-riding issue, so sidelining the demand for divisive policies advocated by sections. Irish Home Rule was a further instance of this strategy. But there was an added dimension: Gladstone disliked Chamberlain, the more so because of what the Prime Minister saw as his lieutenant's disruptive conduct while a member of the 1880-85 government. Gladstone sought to stop Chamberlain replacing him, and so hoped to diminish his stature. Chamberlain had advocated a fairly diluted version of Home Rule, known as the "Central Boards" proposal, which involved conferring greater powers on Irish local government. Gladstone calculated that Chamberlain would be humiliated by having to jettison his pet scheme, and instead follow Gladstone's policy.

Gladstone had made two assumptions which did not materialise, and as a result he had to introduce his proposals in less propitious circumstances than he had envisaged, and with catastrophic consequences. Firstly, he anticipated that the Liberals would win the general election of 1885 with a commanding majority. This would enable him to draft Home Rule legislation without seeming to be acting under pressure from the Irish nationalists. It would also permit him to contemplate a certain

Figure 8. Lord Salisbury, The 3rd Marquess of Salisbury (1830-1903), Prime Minister from 1885 to Jan 1886, July 1886 to 1892, and 1895 to 1902.

Figure 9. Arthur Balfour (1848-1930), nephew of Salisbury, and he succeeded him as Prime Minister in 1902, serving until 1905. He resigned as Leader of the Conservatives in 1911, but remained prominent in politics through most of the 1920s. As Foreign Secretary in 1917 he issued the Balfour Declaration, promising a homeland for the Jews.

degree of dissidence within his party with equanimity, as the rebels were not likely to be numerous enough to defeat his scheme. Secondly, he recognised the need for a period of time in which the party could be brought round to appreciate why he had opted for Home Rule, and a large majority would mean that he did not have to rush precipitately to legislate. The 1885 general election, however, did not produce the result which Gladstone expected. Instead, the balance of power in the Commons lay with the Parnellites: there were 335 Liberal MPs, 249 Conservatives and 86 Home Rulers. This meant that Irish Home Rule would have to be dealt with very early in the life of a Liberal government, in order both to appease the Irish nationalists, who could now block all other legislation, and also to remove them expeditiously from Westminster. When Lord Salisbury's short-lived minority Conservative administration, formed after the election, collapsed early in 1886, Gladstone took office aware that there could be no protracted period in which to educate his party on the case for Home Rule.

The reception of the Prime Minister's scheme among his party at all levels was deeply divided. A large proportion – perhaps one-third – refused to accept this radical switch. At the top echelon of the Liberal party, the response was devastating. The two main contenders for the succession to Gladstone, viz., Lord Hartington (the Whig leader, figure 6) and Chamberlain both declined to follow Gladstone. Lower down the ladder, MPs and constituency activists also opposed the Home Rule proposals. These dissident Liberals were termed Liberal Unionists, and almost immediately organised to defeat Gladstone's scheme. It is important to stress that they still – initially, at least – regarded themselves as Liberals, and emphasised that they had no ideological truck with the Conservatives, even if for tactical reasons they felt obliged to co-operate with the Tories to defeat Irish Home Rule. Liberal Unionists believed that the strength of their opposition to the measure would compel Gladstone to abandon it, whereupon the reunion of the Liberal party would take place. In fact, Gladstone and all subsequent Liberal leaders remained committed to the policy. As a result, Liberal Unionists supported the Conservatives governments of 1886-92 and 1895-1905, with several, notably Hartington and Chamberlain, latterly holding ministerial office under the Tory Prime Ministers, Lord Salisbury (figure 8) and Arthur Balfour (figure 9). Gladstone's second Home Rule, introduced when he returned to office in 1892, reinforced Liberal Unionists' hostility to their old party.

Figure 10. George Darwin (1845-1912), son of Charles Darwin, physicist and protégé of Kelvin, best known for his work on tidal forces

Figure 11. Alexander Crum (1828-93), brother-in-law of Kelvin. He headed the family calico printing firm, built a model village for workers at Thornliebank, and was Liberal MP for Renfrewshire from 1880 to 1885, resigning over Home Rule.

In Glasgow, as elsewhere across Britain, a solid body of Liberal Unionists quickly emerged [3]. Within a month or so, a formal entity had been created, subsequently called the West of Scotland Liberal Unionist Association [WSLUA], to propagandise and organise in support of the unity of Britain and Ireland, and to resist Gladstonian Home Rule. At the centre of this movement was Lord Kelvin [4]. He was involved in the cause from the outset; he exploded to Charles Darwin's son, George (figure 10), soon after the bills were revealed, "Was such madly mischievous legislation ever proposed in the history of England [sic] before?" He was equally scathing about Gladstone's second Home Rule bill of 1892, damning it as "the most monstrous of all bills that ever were put before any

Journal of Physics: Conference Series **158** (2009) 012004 doi:10.1088/1742-6596/158/1/012004

legislative assembly in the world." His untiring commitment was acknowledged by a Glasgow periodical in 1892. There he was acclaimed as:

> "the most prominent Unionist in the West of Scotland. For the past five years no Unionist meeting in Glasgow has been complete without Sir William's [i.e., Kelvin] presence. Whoever has flagged, whoever has fallen by the way, Sir William has known no abatement of enthusiasm. To-day he is as eager in his opposition to the dismemberment of the Empire as he was on the famous 8th of April 1886 [5]."

Kelvin became the first chair of the WSLUA Executive Committee, serving until 1890. Thereafter he held the honorific position of President of the association until his death in 1907. There are several themes which the rest of this paper will explore: firstly, why Kelvin became a Liberal Unionist; secondly, why he was chosen to hold such important positions; thirdly, what contribution he made to the growth of Liberal Unionism; and fourthly, how far his political position changed over two decades

2. Kelvin: from Liberal to Liberal Unionist

By 1886, Kelvin was a long-standing and quite active Liberal. His name appeared in the lists of supporters of party candidates at every general election from 1865 until 1885. Indeed, in the 1874 contest, his brother-in-law, Alexander Crum of Thornliebank (figure 11), unsuccessfully stood for the city as a Liberal. In 1874 and 1880, Kelvin was part of the platform party at the Liberal candidates' eve of poll rallies. His wife reported that he was "in great spirits" at the outcome of the 1880 general election, which resulted in a heavy win for the Liberals [6].

Moreover, Kelvin considered standing for the Liberals at the 1869 election in the newly-created constituency of Aberdeen and Glasgow Universities. The electorate consisted solely of graduates of these two institutions, and as Glasgow was substantially the larger, he would have stood a very good chance of victory – indeed, the Liberals had won the seat in 1868. However, Kelvin declined, stating that he had consulted a sitting Glasgow MP, and concluded that the time and work which would be required meant that he would not be able to continue his teaching and research activities. Cynics might also wonder if he was unwilling to contemplate the substantial loss of income which would arise from abandoning his very lucrative private work, since MPs were then unpaid. Within a decade of deciding against a parliamentary career, he was able to buy a 2,000 acre estate in north Ayrshire, a quite expensive part of the country [7].

Figure 12
Archibald Smith

Figure 13
James Parker-Smith

Instead, in 1868, along with a clutch of Glasgow professors, he energetically promoted the candidacy for the Universities seat of Archibald Smith (figure 12). A distinguished mathematician, Smith had been Senior Wrangler in 1836 at Cambridge, a co-founder of the highly prestigious *Cambridge Mathematical Journal* and was an FRS. The two men had been friends since they had met at Cambridge when Kelvin was a student; Smith was already a Fellow of Trinity College. Moreover, in 1851 Kelvin had proposed marriage on three separate occasions to Smith's sister, Sabina, who rejected him. Smith's son, James Parker-Smith (figure 13), was also to work closely with Kelvin in the cause of Liberal Unionism. Whereas in the city seats Kelvin was only a middle rank Liberal, he took a pivotal role in the universities seat. He acted as the chair of Smith's campaign committee, while the Principal [i.e., Vice-Chancellor] of Glasgow University and the city's most renowned Church of Scotland minister were merely vice-chairs, along with a clutch of professors[8].

Smith was, however, shunted aside to allow a machine politician to have a safe seat. But when the representation fell vacant in 1869, he was selected as the Liberal candidate. Kelvin again played a very prominent role in his campaign. In the midst of the contest, Archie Smith informed his wife that Kelvin was sitting beside him, writing to his many contacts, urging them to support Smith. Despite Kelvin's exertions, Smith lost, and from then until its abolition in 1918, the seat was never again held by a Liberal. Nevertheless, it appears that in 1884 Kelvin was again being mooted as a possible candidate [9].

From the very first intimation of Gladstone's Home Rule plans, Kelvin was in the Liberal Unionist front-line in Glasgow. He was among a small group of dissidents – probably no more initially than nine or ten - who met several times a week in April and May in the offices of A.B. McGrigor (figure 14), the senior partner in one of Glasgow's top legal firms. Here they planned the organisational and propaganda responses to prevent Home Rule. Many of these men, including Kelvin, were voters in Glasgow College constituency, which contained – besides, obviously, the university quarter - some of the most socially exclusive city districts, such as Park and Kelvinside. Six of the original nine plotters were members of the Chamber of Commerce, while McGrigor sat on Glasgow University Court. Hence this mix of academics, prominent businessmen and professional men had pre-existing personal, business and political contacts, so smoothing the way forward.

Figure 14
Alexander B McGrigor, a leading city lawyer and staunch opponent of disestablishment. The initial meeting of the Glasgow Liberal Unionists were held in his offices and he remained very powerful in the party.

Figure 15
David Murray, like McGrigor, was a prominent Glasgow lawyer and extremely active in opposing disestablishment. He participated with Kelvin, McGrigor and Ramsay in the formation of the Liberal Unionist party in Glasgow.

Journal of Physics: Conference Series **158** (2009) 012004 doi:10.1088/1742-6596/158/1/012004

In his first public address as a Liberal Unionist, delivered at the formal launch of the movement in Glasgow, Kelvin defined his main argument in a cry of rage and pain:

> "As an Irishman, I protest with my whole heart and soul [loud cheers] at this insulting proposal of degradation.... We Irish have every reason to demand that you shall not cut us adrift ...[Great Britain] is bound in all honour and duty to listen to us..."

Several themes emerge here, and they can be reinforced and amplified by his later speeches. It is very significant that Kelvin defined himself as an Irishman, something he did repeatedly in subsequent years, indicating a careful choice of terms. He never, so far as I have detected, called himself an Ulsterman in his speeches on the topic; in other words, he did not see the issue of the province's position under Home Rule as a major reason for rejection. Linked to this, he did not particularly emphasise religion as a factor. There were, of course, extremely close links between Ulster and Scottish Presbyterians, and delegations representing the former came to Scotland in the early summer of 1886 to put the religious - or, more correctly, sectarian - case against Home Rule. Next to Ulster itself, Scotland had the greatest number of Orange lodges in the world, and most of these were concentrated in Glasgow and the western area, so it might be expected that this would also be a factor used in Liberal Unionist propaganda. On the whole, however, the Orange or Protestant card was little played by the dissident Liberals, and it was rather the Scottish Conservatives who pushed this line. Orangeism was not seen in Scotland as very respectable, as its membership was overwhelmingly working-class [10].

His more specific arguments can be classified into five categories. Firstly, Home Rule, so far from being beneficial to Ireland, would in reality prove very detrimental. Kelvin contended that the pursuit of independence was a "chimera", and the Irish should instead concentrate on working to improve the whole United Kingdom, and to promote Irish interests within that framework. He believed that "my beloved Ireland shall be happy and prosperous in its natural position as a member of the United Kingdom." Furthermore he commented that the measure would do grave damage to Ireland, somewhat less to Scotland, and considerably less to England [11].

Secondly, Kelvin argued that the economic prosperity of Ireland would be harmed, not increased, by Home Rule. The counterpart to this was the view that the economy of the Glasgow region would be gravely damaged by Home Rule, for there were numerous manufacturing, commercial and financial connections between the west of Scotland and Ireland – and particularly, but not exclusively, Ulster. Within a fortnight of the publication of the bills, the Glasgow Chamber of Commerce met and passed resolutions condemning the proposals as inimical to trade, investment and industry in both Ireland and Scotland. The chamber, the oldest in Britain, contained virtually all the city's leading businessmen, and Kelvin was himself a member. The main proposition was put by A.B. McGrigor, and although it is not clear whether Kelvin attended, it is virtually certain that he would be in agreement with this analysis [12].

Thirdly, social stability and maintenance of law and order throughout Ireland would be seriously endangered by self-government. This threat arose from two different factors. On the one hand, as he stated in 1892, the impossibility of Belfast accepting Home Rule was "manifest" to all, and the bloodshed in the city in 1886 when the mere threat of it being imposed indicated that civil war would break out if it were ever carried into law. On the other hand, an Irish parliament could not be relied on to suppress criminality and violence. Clearly Kelvin was implying collusion between nationalist politicians and the agrarian and Fenian violence. He claimed that the collapse of civil authority would mean that "the law-abiding industrious people who had made the prosperity of Ireland could not accept the position" [13].

Fourthly, Home Rule would provoke a profound constitutional crisis. It would prove impossible to stop an Irish parliament from extending its powers, as effective supervision from Westminster would be difficult. This, of course, would destroy the sovereignty of the UK parliament. Somewhat contradictorily, he claimed that the MPs sitting in Dublin would be impotent to influence legislation in

Journal of Physics: Conference Series **158** (2009) 012004 doi:10.1088/1742-6596/158/1/012004

the British parliament, merely "being allowed to play at parliamentary debating in a Parliament House of [their] own." Further friction would arise because the Irish people, with no presence in Westminster, would be unable to influence imperial, foreign and defence policy matters, although they would be contributing through taxation to the costs of these. It would also be wrong if Westminster were to ignore the demand of one-sixth of the Irish people to retain the union. But there was for Kelvin another defect in a separate Dublin parliament. Scotland and England would thereby be deprived of the positive gains which Irish MPs had made in Westminster to the better governance of Britain. They had done "a great deal of good" there. Furthermore, the rest of the United Kingdom was right to seek the help of the best Irishmen to run the united country, as there was a need for all available men of ability to participate in this task [14].

Lastly, at least initially, Kelvin deplored Gladstone's land purchase scheme as a burden on Scottish and English taxpayers. He felt that the "only remedy possible, and all the remedy that is needed" to solve the land question was for the government to offer "moderate encouragement and help towards business habits and arrangements" between landowners and the tenantry [15].

Of course, such complete rejection of Gladstone's proposals required that Kelvin should offer an alternative approach to resolving the Irish problem. Over the decade or so after 1886, he did put forward a sort of coherent vision. The constitutional difficulty was to be overcome by offering the fullest degree of local self-government consistent with retaining the overriding sovereignty of Westminster. This was really akin to the relationship which colonies such as Australia, Canada and New Zealand had with Britain. Soon after 1886, he became an ardent champion of land purchase, now seeing this as a key stage in pacifying agrarian protest. He also insisted on the restoration of law and order throughout Ireland as essential to create the groundwork for economic and social development. But he emphasised in 1886 that the whole process would be gradual and long-term: "Let us persevere - persevere even with small hope of seeing the fruits..", in the expectation that 'perseverance in well-doing ...shall bear fruit in the happiness and prosperity of those that come after us, although we may not live to see it.' A few years later, he claimed that the policies of the Conservative government, with Liberal Unionist backing, had rescued Ireland from anarchy and were setting the island on the road to prosperity, so completely vindicating the stance taken by Liberal Unionists [16].

Figure 16
Professor George Gilbert Ramsay, who succeeded his uncle as professor of Humanity [Latin] at Glasgow University, and occupied the chair for 43 years – he was Dean of Faculties in his 80s! He was the most active university Liberal before 1886, and thereafter a staunch Liberal Unionist

Figure 17
Ramsay was a very eminent classical scholar, his most important work being his edition of Juvenal and Persius

3. The choice of Kelvin as the leader of Liberal Unionism in the West of Scotland

The choice of Kelvin to be the first chair of the WSLUA Executive Committee was probably the outcome of a number of factors. Partly it was who he was, but perhaps equally important, it was who he was not. *Prima facie*, he was not the most obvious or natural candidate for the post. Although, as noted above, a long-standing Liberal, he had not been in the city party's front rank, neither serving on any candidate's central (i.e., executive) committee, nor holding any position in the Glasgow Liberal Association itself. In the College constituency, he was not the head of the Liberal party, and even among the professoriate he was not the foremost Liberal. That honour belonged to G. G. Ramsay (figures 16 and 17), as was evidenced in the 1880 election. At a Liberal rally at which both Kelvin and Ramsay sat on the platform, the end of the meeting was delayed by repeated shouts from the audience for the latter to speak, and it was only after he had addressed them that they left.

But Kelvin had important positive qualities. Firstly, he was arguably Glasgow's only citizen with an international reputation, so he added lustre and renown to the Liberal Unionists. Secondly, his acclaimed status as one of, if not the, foremost British scientists of the day imparted an aura of rationalism and dispassionate analysis, rather than mere prejudice and partisan adherence, to the cause. His long career – some forty years – at the university meant that a very large of number of students had been taught by him (Physics was then a central component of the Scottish degree curriculum), and others would have certainly heard of him. Additionally, his career as a highly successful entrepreneur would have made him acceptable to hard-headed businessmen, who might otherwise have regarded him as a typical academic inhabitant of an ivory tower.

As implied above, it was also very useful that he had not been a leading local activist, because he was not closely identified with different camps and hotly disputed causes, in the way that more prominent Liberals were. One obvious choice would have been to put an aristocrat at the head: late 19th century Britain was still a highly deferential society, and someone like the Duke of Argyll, a former Liberal cabinet minister and owner of vast estates in the nearby eponymous county, was a clear option. But two intersecting forces had to be borne in mind by the WSLUA leadership. One was that two specifically Scottish questions had dominated the 1885 general election, fomenting much internal Liberal feuding. The other was the paramount need to retain within the Liberal Unionist embrace both the moderate Whig and the radical Chamberlainite wings [17].

One of these Scottish issues in 1885 was the plight of Highland crofters [18]. This class, the equivalent of the small-holding peasantry of the west and south of Ireland, faced a major threat to their existence in the middle of the 1880s. Changes in land use strategies by the great estates, which involved ending sheep-farming and moving instead to developing sporting pursuits, threatened the marginal subsistence regime of the crofters. Their customary practice of putting livestock on to higher land for summer grazing was in conflict with deer-stalking and grouse-shooting, which maximised estate revenues. With no paper legal rights, many crofters faced eviction or emigration. Unlike the earlier phase of Highland clearances between c1815 and c1850, the crofters resisted with violence in the so-called "Crofters' War" (Figures 18 and 19) which began in 1881-2. An ensuing Royal Commission into their grievances gathered eloquent testimony to the sense of injustice and maltreatment experienced by the Highland peasantry.

The public perception of landowners as a grasping and brutal class created a spirit of great hostility which was not just confined to the Highland region, but was felt across all of Scotland. This sentiment was particularly forcefully voiced in Glasgow, and the nearby western Lowland area. Partly this was because large numbers of Highlanders had settled in these places: Glasgow itself had a sizeable Highland community. Additionally, many Lowland radicals were opposed to single ownership of large landed estates as a form of monopoly which affronted their free trade ideology. When the Third Reform Act of 1884 gave the vote for the first time to the Highland peasantry, they formed the Crofters' party, and in the 1885 election contested all the Highland seats, opposing landowner candidates, whether Conservative or Liberal. The Crofters' case aroused great demonstrations of sympathy in cities like Glasgow; mass meetings were held, funds were raised, and Lowland radical Liberal candidates pledged to support the crofters' campaign for legislation on the lines of the Irish Land Act of 1881. The result of the elections was a clean sweep in the Highlands for the Crofters' candidates.

Figure 18
The "Crofters' War"

Figure 19
Monument to the "Crofters' War"

Most Whiggish Liberal Unionists felt very little identification with the crofters, and the more rigorously ideological among them were alarmed at the breach of private property rights which the crofters' demands required. But the absolute necessity of maintaining the broadest possible phalanx of anti-Home Rule Liberal support suggested that anyone identified with landownership, such as the Duke of Argyll (who had been a vociferous opponent of land reform in both Ireland and the Highlands), would be unpalatable to Radical Liberal Unionists.

The second controversy to plague Scottish Liberals in the 1885 election was the campaign to disestablish the Church of Scotland [19]. It should be borne in mind that, unlike England, Wales and Ireland, the established church in Scotland was – and still is – Presbyterian, not Episcopalian. The main advocates of Scottish church disestablishment were two other Scottish Presbyterian churches, viz., the Free Church of Scotland (which Kelvin attended when at his Largs estate) and the United Presbyterian Church. The combined membership of these two denominations comfortably exceeded that of the state Church of Scotland – hence the bitterness of the debate. The Radicals strove to have this plank endorsed by Liberal candidates, but a significant body within the party, named the Church Liberals, opposed this. Many Whigs who became Liberal Unionists were very closely identified with the latter. A.B. McGrigor, Professor Ramsay and David Murray (figure 15), a partner in a large city legal firm, formed a local Church Defence Society to counter the disestablishers, who were gaining control of the whole Glasgow party. This society held a major rally in Glasgow at the height of the 1885 election campaign, at which ten university professors were either present or sent apologies - but Kelvin was not among their number. Three future prominent Liberal Unionists, McGrigor, Ramsay and W.V. Jackson, spoke at the meeting. The state church issue was the cause of bitter internal feuding in a number of seats, and the selection of anti-state church candidates very nearly broke the Liberal party in two constituencies which became Liberal Unionist strongholds. In College, the President of the constituency Liberal Association announced that he would not vote for the pro-disestablishment candidate, and he was supported by McGrigor and Ramsay. In Central, W.V. Jackson temporarily resigned his party membership in protest at the choice of a disestablisher as candidate. The Radical faction retaliated by denouncing the Church Defence Society as a covert Tory body, and there were calls for its members to be expelled. The dispute did not subside after the election, rather it was intensified, since all seven Glasgow Liberal MPs were hostile to the state-church link. At the start of 1886, the Church Liberals were plotting to form a permanent organisation to fight the disestablishment wing in the constituencies. McGrigor, Ramsay and Murray spearheaded this movement in Glasgow, and they were in communication with the Edinburgh Church Liberals [20].

Journal of Physics: Conference Series **158** (2009) 012004 doi:10.1088/1742-6596/158/1/012004

The question of Irish Home Rule, however, cut horizontally across the two camps. Prominent disestablishers like Alexander Cross and A. Craig Sellar, MP for Partick, advocated retention of the union, as did McGrigor and Ramsay. Hence to choose any of these men, who had a far more prominent profile in the Liberal party than Kelvin, might have resulted in the other side declining to remain in the Liberal Unionist big tent, thereby enhancing the prospect of Gladstonian Home Rule triumphing. Kelvin's ecclesiastical commitments can be described as flexible. His father had been a Presbyterian, but as a student at Cambridge University, Kelvin worshipped as an Anglican, yet when at his professorial house in Glasgow University, he attended the university chapel, which was attached to the Church of Scotland. However, when staying at his Largs estate, he went to the local Free Church. Although in theological terms, there was little to differentiate these two Presbyterian sects, there was then a profound enmity between them, so Kelvin's latitudinariansim is quite striking. He seemingly went to Largs Free Church mainly because the minister was the brother-in-law of his first wife. However, he stopped going there over Home Rule, and instead attended the town's Episcopalian church, so that there was the delicious paradox of Kelvin being a Presbyterian in Glasgow, but, thirty miles down the coast, he was an Anglican [21].

But for whatever reason Kelvin was chosen, his appointment proved immensely beneficial to the Liberal Unionists. He evidently had the inestimable quality of being exceptionally adept at chairing meetings in a manner which prevented differences of opinion from escalating into serious rows or splits. This was a vital element in a body with many fissiparous proclivities. No doubt his gravitas and eminence contributed substantially to ensuring a degree of harmony, but he also seems to have had a naturally emollient style, and an instinct for finding consensual ground. These facets were most clearly displayed when he chaired the association's three annual general meetings from 1903 to 1905. The party across Britain was deeply split over Chamberlain's tariff reform (i.e., protectionist) proposals. The Duke of Devonshire (as Hartington had now become) resigned as president of the party in protest, and in the Glasgow area, too, there were many resignations and even departures to the Liberals. Kelvin managed to conduct the business of the meetings in such a way that completely opposing resolutions relating to tariff reform were dealt with without provoking a schism, for which he received fulsome praise from sympathetic newspapers. This even-handedness on Kelvin's part, which was acknowledged by all sides, was the more remarkable because, as discussed later, he was an early and ardent advocate of tariff reform [22].

4. The West of Scotland Liberal Unionist Party under Kelvin

The West of Scotland Liberal Unionist Association rapidly earned the reputation of being one of the two most efficient regions of the party, along with Durham [23]. Three days after the First Reading of the bill, a public meeting was called, and this rapid response was well ahead of any other part of Britain. Once the association acquired a formal structure in early May, the Executive Committee became almost hyperactive, with several meetings a week taking place throughout that month. Sub-committees were promptly established to ensure speed of action in areas of priority action.

One urgent objective was to set up a rival organisation to the Home Rule Liberal party. Branches were established in constituencies across the region, and by mid-June all seven Glasgow seats had Liberal Unionist associations. Recruitment of members was prosecuted vigorously from the start, and by July, over 2,000 had joined the breakaway party. £400 was raised locally within two months, so facilitating the funding of activities. At the end of June, the association acquired office premises in central Glasgow, and administrative staff were appointed. The WSLUA also immediately established links with the national party organisation, based in London, and it sent six delegates to that body's earliest meetings. Kelvin served as a WSLUA representative on the national (British) Liberal Unionist Council for many years, often arranging Royal Society business trips to coincide with the council's meetings. A further stage in organisation was the formation in 1889 of the WSLUA women's section, whose inaugural meeting both Kelvin and his wife attended [24].

A second central preoccupation was to keep Liberal Unionists committed to the cause, and to recruit new adherents. As the first leaflet produced by the association put it, a key objective was to bring together all anti-Home Rule Liberalism, and to "create, inform and express" this opinion in the constituencies. A great deal of effort was put into these propaganda aspects. Both Chamberlain and Hartington, as well as many second tier Liberal Unionist politicians, came to speak at mass rallies. This reinforced the sense that Liberal Unionists in Gladstonian strongholds were not isolated or impotent in the face of the local prevailing current of Liberalism. Meetings were also held across the region to initiate and sustain local branches. Pamphlets advocating the Liberal Unionist standpoint were written and published for distribution.

A third area of concern was to form an understanding with the Conservatives in order to avoid the duplication of candidates in a seat standing at elections as Unionists, lest a Gladstonian Liberal slip in on a minority vote. This issue was addressed from the very start, and by the 1886 election, all local difficulties had been resolved, so that only one Unionist candidate came forward in every seat in the association's area. This sub-committee continued to be very active throughout the WSLUA's existence, as local disputes between the two parties were a recurring event.

The best gauge of the efficacy of the WSLUA is probably to be found in general election results, as shown in Table 1. Clearly, compared to England and Wales, the West of Scotland section performed significantly more successfully. The general election of 1900 was the high-water mark for the Unionists in Scotland, especially in Glasgow itself, where all seven seats were won, three of them by the Liberal Unionists, their highest tally for the city. While, strictly speaking, only the first general election in 1886 fell within Kelvin's term as chair of the WSLUA Executive Committee, it is not unreasonable to surmise that the foundations for the party's continuing strong performance were laid in the early phase. This poses the question as to how far Kelvin should be personally credited with these achievements.

Table 1: Liberal Unionist MPs Returned at General Elections, 1886-1900

	England & Wales		West Scotland	
Election	No	%	No	%
1886	59	12	8	28
1892	30	6	6	21
1895	52	11	6	21
1900	51	11	7	24

Note: The percentages relate to the total number of seats in each area.

It is quite interesting to look at the results in the Glasgow College constituency, where Kelvin voted. In 1886 and 1892, Charles Cameron (figure 20) was returned. He was born in Dublin and had qualified as a doctor, but ran a Glasgow newspaper. He had been elected as a radical Liberal MP for Glasgow in 1874, was a keen advocate of both disestablishment and the Crofters' campaign, and supported Gladstone over Home Rule. In 1892, Sir John Stirling-Maxwell (figure 21), a wealthy landowner with estates near Glasgow and in Inverness-shire, was the beaten Conservative candidate, Kelvin having proposed his adoption. Sir John returned in 1895 to take the seat

Kelvin took a leading part in a number of aspects of organisation, notably speaking at meetings and electioneering. In June 1886, he travelled to Greenock to put the Liberal Unionist case, and in October he attended a Belfast rally addressed by Chamberlain. In April, he went to Edinburgh with fellow Glasgow Liberal Unionists to hear Hartington and G.J. Goschen (figure 22), another major anti-Home Rule minister, speak at huge meetings. In the 1886 contest, he spoke regularly on behalf of Unionist (both Conservative and Liberal) candidates across the west of Scotland, including going on his beloved yacht, the Lalla Rookh, to Oban to support the Tory candidate for Argyll. He was again active in the 1892 election, and also participated in by-elections, including speaking in 1890 at the

Journal of Physics: Conference Series **158** (2009) 012004 doi:10.1088/1742-6596/158/1/012004

Partick contest, where the Liberal Unionist candidate was James Parker-Smith. Even after he stopped being chair of the Executive Committee and assumed the less active position of President, he retained an interest in the detail of party machinery. In autumn 1890 he invited an influential executive member to come down to Largs to discuss organisational matters [25].

Figure 20
Charles Cameron
Elected MP for Glasgow College
in 1886 and 1892

Figure 21
Sir John Stirling-Maxwell
Elected MP for Glasgow College
In 1895

On the other hand, Kelvin was absent from a considerable number of Executive meetings during his period as chair: out of a total of about 35, he attended only 13, or just over one-third. It must be remembered that there were extenuating circumstances. He was, after all, a very busy individual. He was frequently away in London, attending Royal Society meetings and committees. Furthermore, his business affairs, scientific research and teaching duties took up a great deal of his time. He continued to hold the chair of Natural Philosophy (i.e., Physics) until 1899. At the height of the Home Rule crisis, in May 1886, he was receiving a veritable stream of letters from colleagues and research assistants on various scientific topics: one particularly pressing matter concerned problems with compasses. All of this diverse range of activities – plus travelling to and from his Largs home, some thirty miles from Glasgow - is quite remarkable for a man entering his 62nd year in 1886.

Figure 22. George Goschen (1831-1907), who joined the Liberal Unionists in 1886, and served under Lord Salisbury as Chancellor of the Exchequer. He could not work with Chamberlain and joined the Conservative party in 1893. The cartoon shows him "ascertaining which way the wind blows".

Journal of Physics: Conference Series **158** (2009) 012004 doi:10.1088/1742-6596/158/1/012004

It is pretty clear that most of the regular work of the WSLUA was conducted in his absence by others, who were all highly competent. A.B. McGrigor normally chaired Executive Committee meetings when Kelvin was away, and he was undoubtedly the *eminence grise* of the association, sitting on most of the key sub-committees. He was especially active on the committee which liaised with the Conservatives over the selection of mutually acceptable candidates, the most delicate and important of tasks. When Kelvin gave up the chair, McGrigor succeeded him, but resigned within a year, possibly due to ill-health. The day to day running of the association was the responsibility of the secretary, Robert Bird. Bird was evidently highly proficient in his job, and seems to have initiated many of the organisational techniques adopted by the WSLUA. This is seen in two accounts he wrote in 1887 and 1889 for the national party's journal, wherein he outlined the methods which he had used to build up the association's effectiveness [26].

It may be appropriate here to discuss the reasons why a peerage was conferred on Kelvin in 1892. At the time it was acclaimed widely as a fully justified acknowledgement of his massive and wide-ranging contributions to both theoretical and practical science. But the authors of the weightiest modern study of Kelvin's life argue that the title was at least in part a recognition of his political services to the cause of Unionism, and there are very good grounds for agreeing with this verdict. Firstly, giving peerages to scientists was virtually unknown before then, and it was many years before the next case. Secondly, awarding honours was part of the means by which the Liberal Unionist leadership (like the other parties' leaders) rewarded supporters, and in this the Conservative governments collaborated out of obvious self-interest, So successful were the Liberal Unionists that promnent Scottish Conservatives complained about what they viewed as excessive preferential treatment accorded to the former. Indeed, in the same year as Kelvin's ennoblement, a peerage was also bestowed on a prominent Glasgow Conservative, presumably to appease the disgruntled Scottish Tories. It is highly significant that Kelvin's next but one successor as chair of the WSLUA Executive Committee was made a baronet in 1903. Mathew Arthur was a wealthy businessman, but not otherwise a particularly significant figure, so it should best be seen as at least in part a political award [27].

It seems clear, however, that Kelvin did not acquire his peerage in return for substantial financial donations to the party. In a letter of 1892 sent by Hartington to Lord Salisbury, some men identified as candidates for an honour were listed as having contributed generously to the Liberal Unionist coffers, including a Glasgow businessman who had "collected" £20,000. Kelvin by contrast made only quite small donations. At the launch of the WSLUA, he gave £20 to its guarantee fund, which raised £400 from some 25 individuals. Again, the association's special election fund for the 1895 general election totalled £27,375. Kelvin was one of eight who gave £100, while seven contributed higher amounts, with £2,000 being the largest donation [28].

It is highly unlikely that Kelvin threw himself so strenuously into the Liberal Unionist cause with the calculation in mind that a title might eventually be forthcoming. Six months after the crisis began, he expressed very clearly the sense of public obligation which had impelled him to be active: "Much as any of them disliked leaving their own affairs and entering upon a political contest, none of them would shirk so long as there was any idea of breaking up the legislative union of the United Kingdom." At the same time, moreover, he confidently predicted that: 'the disruption was a temporary disease…he had not the slightest doubt that there would be a reunion of the Liberal party [Hear, hear]'. He persisted for some years with this line of argument, stating that reunion was perfectly possible if only the Gladstonian Liberals would abandon Irish Home Rule. Furthermore, the rowdy reception which he sometimes endured at meetings in the fevered atmosphere of 1886 strongly suggests that Kelvin was acting out of true conviction, rather than seeking a title. For instance, in June he chaired an election meeting in a mainly Gladstonian working class district of the College seat. As Kelvin rose to make some introductory remarks, he was met with "a storm of hisses and groans which lasted for some time." He then claimed that Home Rule would be bad for England and Scotland, to which there were cries of "no, no", and "sit down", and he was heckled persistently [29].

5. Kelvin's shifting political opinions, 1886-1908

The rupture of 1886 and subsequent developments led Kelvin to reappraise his political views in a number of areas over the last twenty years of his life. As noted above, initially he highlighted his continuing Liberal beliefs, and his expectation that the rift would be temporary. However, about two years after the schism, his position became more ambiguous. One aspect of this equivocation was to assert that the Gladstonian Liberals were not true Liberals. In 1896 he referred to the "ex-Liberal party" and the "so-called Liberal" party. The following year he returned to this: "All good and true Liberals ought to belong to that organisation [the Liberal Unionists], and he believed that all good and true liberals did belong to it." [30]

If he was indeed a Liberal, clearly Kelvin needed to justify his unstinting support for the Conservative governments of 1886-92 and 1895-1905. He did this by asserting that, thanks to the influence of his party, Liberalism had suffused Conservative policies. In October 1888, he proudly acclaimed the work of the Tory administration:

> "What government at any time in the last thirty years had brought forward more truly Liberal measures and had carried into law more truly Liberal acts of parliament than the present government?.. They had acted in harmony with the Liberal [i.e., Liberal Unionist] party, and they had produced Liberal measures (Cheers)."

He maintained this viewpoint for the rest of his life, contending in 1903 that "there was a strong force of Liberalism in the [Unionist] government" [31].

Virtually simultaneously, Kelvin elaborated a quite different, and rather contradictory, analysis. He became wedded to the idea that government by party was disappearing, to be replaced by a kind of consensus of technocrats and well-disposed individuals. Intriguingly, here he often invoked Darwinian evolutionary concepts to justify his thesis that politics and society were advancing to a new, higher stage of development. So, in 1888, he asserted:

> "Darwin, in stating his doctrine of natural selection, spoke of permanent and non-permanent species. Did anyone imagine that government by party was a permanent species? It had served its term and was played out. We were living under a government not by party…The political millennium, he was satisfied, had come. We had reached the fourth stage – government by common sense."

Given that Kelvin harboured doubts about the overall validity of Darwin's theory, his invocation of part at least of it is fascinating [32].

A further twist to this post-party argument was posited from 1900. He began to propose that parliament should operate on the lines of "a well-conducted limited liability company", with a broad base of supportive MPs giving virtually uniform endorsement to the actions of the government of the day, just as shareholders would stand behind the directors. Dissidence would be confined to a small minority, and would be only occasional, rather than organised and systematic. Politicians, he explained, had an ideal of parliament as consisting of 350 MPs on one side and 300 on the other, whereas 650 MPs all on one side was the ideal of "those who, like himself, were not practical politicians." Again, he invoked Darwin: "Natural progress, natural selection, evolution seemed to him more likely to produce a rational system…" [33]

His efforts to defend the Liberal Unionists' support for the Conservatives led him to become increasingly sympathetic to the latter's policies in areas other than Ireland. By around 1900, Kelvin seemed on the key political issues of the times to be increasingly detached from the Liberal tradition. During the Boer War of 1899-1902, he was an unstinting champion of the Conservative government's assault on the Boers, although most Liberals regarded the war as a flagrant breach of their party's most cherished foreign and imperial policy principles, especially disliking the atmosphere of hysterical jingoism which surrounded it. Kelvin's old Liberal Unionist comrade, Professor Edward Caird (figure

Journal of Physics: Conference Series **158** (2009) 012004 doi:10.1088/1742-6596/158/1/012004

23), now Master of Balliol College, was one who opposed the war. By contrast, Kelvin espoused the cause with imperialistic fervour, insisting in 1901 that the government was: "bent on one object, to do good work in South Africa by establishing firmly and forever the British regime there, and all the good things that British regime meant." It is doubtful whether Kelvin meant that the concentration camps, a British innovation in this war, were part of the "good things"[34].

Figure 23. Edward Caird, Professor of Moral Philosophy at Glasgow from 1866 to 1893, in which year he became Master of Balliol College Oxford. Caird was a Liberal Unionist with Kelvin, but, unlike Kelvin, strongly opposed the Boer War.

Figure 24. Andrew Bonar Law, Leader of the Conservative party from 1911-21, and then briefly, both party leader and Prime Minister from October 1922 to May 1923.

In 1903, as noted above, Joseph Chamberlain launched his Tariff Reform agitation. Free trade had for over seventy years been the shibboleth of the Liberal party, and to most even the mildest infringement was to breach the ultimate taboo. While a considerable number of Glasgow Liberal Unionists – both Whig and radical – vehemently opposed Chamberlain's initiative, Kelvin quickly emerged as a passionately committed protectionist. He may have been influenced by James Parker-Smith, who was the foremost spokesman for the cause in Glasgow. In November 1903, he agreed to become a Vice-President of the Tariff Reform League, remarking that "Now I am more and more convinced that Chamberlain is right." He aligned himself with the most hard-line protectionist stance by insisting that food imports should be not exempt from tariffs. It seems indicative of Kelvin's journey across the political landscape in the two decades from the first Home Rule bill that one of his last public political occasions was to speak at a Protectionist rally held in April 1907 in support of Andrew Bonar Law (figure 24), the future leader of the Tory party [35].

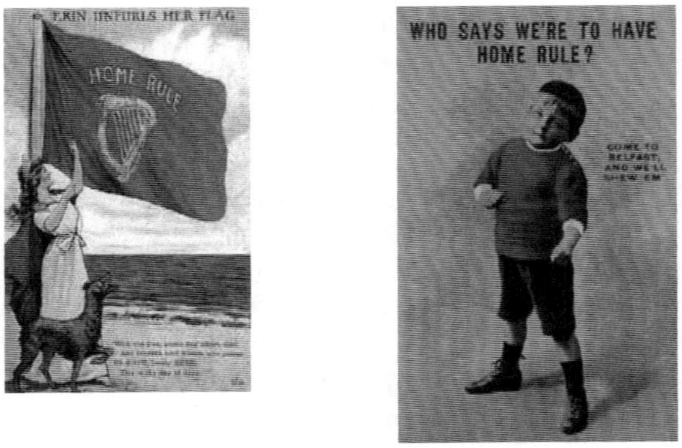

Figure 25. The spectre of Home Rule caused great divisions – and not just in Ireland!

References

[1] For this and the next four paragraphs, see:
Hoppen K T 1998 *The Mid-Victorian Generation, 1846-1886* (Oxford: Oxford University Press) pp 668-90
O'Day A 1996 Home Rule and the historians *The Making of Modern Irish History* ed Boyce D and O'Day A (London: Routledge) pp 141-62
Shannon R T 1998 *Gladstone: Heroic Minister, 1865-1898* (London: Allen Lane) pp 365-441

[2] One of the 86 Home Rule MPs sat for an English seat – T.P. O'Connor held the ironically named Liverpool Scotland constituency as an Irish Nationalist.

[3] For the Scottish and Glasgow position, see:
Burness C 2003 *'Strange Associations': The Irish Question and the Making of Scottish Unionism,1886-1918* (East Linton: Tuckwell Press) pp 44-68
Hutchison I G C 1986 *A Political History of Scotland, 1832-1924* (Edinburgh: John Donald) pp 162-7, 207-12, 225-7
McCaffrey J 1971 The Origins of Liberal Unionism in the West of Scotland *Scottish Historical Review* **50** (1)

[4] At this point, of course, he was still Sir William Thomson, and did not become Lord Kelvin until 1892, but it has seemed simplest to refer to him by the latter name throughout.

[5] Thompson S P 1910 *The Life of William Thomson, Baron Kelvin of Largs* (London: MacMillan), Vol. 2, p. 857; WSLUA M[inute] B[ooks] 28 Nov. 1894. (These are held in the Manuscripts Room of National Library of Scotland, Edinburgh, temporary finding reference: Acc. 10424/19-21); *The Bailie* 13 Jan. 1892

[6] E.g. *Glasgow Herald* 4 Jul. 1865, 2 Feb. 1874; Thompson *Kelvin* vol. 2, p. 758

[7] Thompson *Kelvin* vol. 2, pp. 553-8

[8] Smith C & Wise M N 1989 *Energy & Empire. A Biographical Study of Lord Kelvin* (Cambridge: Cambridge University Press) pp 174-6, 69-70, 141-4

[9] Glasgow Council Archives, TD 1/918 A. Smith to Mrs Smith 13 Oct. 1869; Smith & Wise *Industry and Empire* pp. 803-4

[10] *Glasgow Herald* 22 Apr. 1886

[11] *Glasgow Herald* 22 Apr. 1886

[12] *Government of Ireland Bill. Proceedings in the Chamber of Commerce and Manufacturers in the City of Glasgow on 19 April 1886; Glasgow Herald* 20 Apr. 1886

[13] WSLUA MB 27 Jan. 1892, also 16 Apr. 1890; *Glasgow Herald 1 Jul. 1886*

[14] WSLUA MB 27 Jan. 1892, 30 Dec. 1889; *Glasgow Herald* 22 Apr. 1886; Thompson *Kelvin* Vol. 2, p 920; WSLUA MB 29 June 1886

[15] Thompson *Kelvin* Vol. 2, pp. 855-6

[16] WSLUA MB 26 Nov. 1896; Glasgow *Herald* 22 Apr. 1886; WSLUA MB 3 Dec. 1889, 18 Nov. 1890

[17] E.g. Thompson *Kelvin* p. 861 (Kelvin to Lord Rayleigh 18 Jul. 1886); The *Liberal Unionist* **1** (12) (15 June 1887); WSLUA MB 1 May 1893

[18] For the background, see:
Devine T M 1994 *Clanship to Crofters' War* (Manchester: Manchester University Press)
Fry M 2005 *Wild Scots* (London: John Murray)

[19] The context is to be found in Hutchison *Political History* and McCaffrey Origins of Liberal Unionism (see [3])

[20] *The Church of Scotland. Report of a Meeting of Scottish Laymen of Different Denominations opposed to the Disestablishment and Disendowment of the Church of Scotland. Glasgow 20 October1885*; Glasgow University Library Special Collections Department, Mu44.e20 – correspondence between L. Mackersy and D. Murray, Jan.1886; (RVC) c. Apr 1886 *Scottish Liberals and Disestablishment – A Proposal for a National Liberal Union*

[21] Thompson *Kelvin* Vol 2 pp 1087ff.
[22] WSLUA MB 3 Dec. 1903, 23 Nov. 1904, 29 Nov. 1905; *Scotsman* 3 Dec. 1903
[23] For more information on this and the following three paragraphs, see the works cited in [3]
[24] WSLUA MB 8 Dec. 1891; *The Liberal Unionist* **6** (Feb. 1892)
[25] WSLUA MB 7 June 1886; Thompson *Kelvin* Vol 2 p. 862; *Glasgow Herald* 1 May 1886; Thompson *Kelvin* vol I2 pp 860-1, 911-12; *Glasgow Herald* 11 Feb. 1890; Glasgow Council Archives TD1/357 Kelvin to J.P. Smith 19 Sep. 1890
[26] *The Liberal Unionist*, **1** (12) (15 June 1887), **3** (38) (Mar. 1889)
[27] Smith & Wise *Energy & Empire* pp 799-807
[28] Jenkins TA 1990 The funding of the Liberal Unionist party and the honours system *English Historical Review* **105** (5); WSLUA MB 10 May 1886, 8 Aug. 1900
[29] WSLUA MB 2 Nov. 1886; *Glasgow Herald* 29 June 1886
[30] WSLUA MB 26 Nov. 1896, 10 Nov. 1897, also 1Oct. 1888
[31] WSLUA MB 1 Oct. 1888, 2 Nov. 1903, also 23 Mar. 1888, 16 Apr. 1890, 26 Nov. 1896, 19 Oct. 1898; *Glasgow Herald* 3 Dec. 1889
[32] WSLUA MB 1 Oct. 1888, also 21 Nov. 1900; Smith & Wise *Kelvin* pp. 636-41
[33] WSLUA MB 21 Nov. 1900, 26 Nov. 1901
[34] WSLUA MB 24 Nov. 1899
[35] Glasgow Council Archives TD 1/144 Kelvin to J.P. Smith, 25 Nov. 1903; Thompson *Kelvin* Vol 2, pp 1129-30, 1198

Journal of Physics: Conference Series **158** (2009) 012005 doi:10.1088/1742-6596/158/1/012005

Kelvin's ideal foam structure

Denis Weaire
School of Physics, Trinity College Dublin
Ireland

E-mail: dweaire@tcd.ie

Abstract. Among his many extraordinary accomplishments, Kelvin was a pioneer of crystallography, elasticity and materials science. These interests came together to inspire his speculation on the nature of the ether in 1887. He conceived it to be an ordered liquid foam, of minimal surface area. Kelvin's ideal structure of foam has been realised in the laboratory only recently. In the meantime it was surpassed (in terms of surface area minimisation) by the Weaire-Phelan foam, which is the basis for one of the main buildings of the Beijing Olympics.

1. Kelvin's versatility

The span of Kelvin's published work is enormous. Not only did he publish in many fields of physics, but he cross-fertilised them, often bringing the fundamental and the applied to bear on each other. It is an all-embracing style that few could even aspire to emulate today, even if they had his talent and energy.

His proposal of a structure for the ether of space [1] is a good example and it happens to be topical today. Here he was addressing one of the great fundamental questions of his age: what invisible material fills space, enabling light waves to travel through it? What is the nature of this *ether*? It was the "dark matter" of the 19th century.

It had been thought for half a century or more that the light waves must be elastic waves in a material ether, but ordinary materials do not have the right properties to account for light waves in this way.

Even as Kelvin dwelt upon this paradox, others began to consider it false or misguided. Instead of the elusive ether they accepted the more abstract field equations of Maxwell [2], and the doctrine of the so-called Maxwellians [3]. But Kelvin was not persuaded by that alternative view, and insisted that the ether was "a real thing". He was familiar with many materials that had unusual properties: might one of these not be a candidate for the kind of stuff that made up the ether?

He continued to turn the problem over his mind until, lying in bed in his seaside mansion in 1887, he dreamed up his foam model of the ether. In framing and elaborating this he was to call on his knowledge of elasticity, wave theory and crystallography as well as his general background in materials science.

2. Kelvin and materials

No doubt pure curiosity played a role in Kelvin's interest in materials, but it also derives from the applied research that made much of his reputation and most of his riches.

Beginning with a theoretical study of the propagation of telegraphic signals, he was gradually drawn into a concern for the properties of the very materials that constituted the telegraphic cable. On board the cable ships, struggling with practical realities, he saw the essential role of those materials. If they failed, then so did the entire enterprise.

The story of the gradual improvement and optimal exploitation of those materials is reminiscent of twentieth-century solid state physics. Just as Kelvin was led to demand better characterised materials, so in our own time did the semiconductor industry require improved control of silicon, and arcane research was seen to underpin industrial success. If we trace a thread of continuity between the two periods, Kelvin deserves to be called the first solid state physicist.

The telegraph cable was at that time (and for a long time to come) primarily made up of a copper core and a gutta-percha sheath. They played the role of conductor and insulator respectively, and those properties required to be standardised and optimised. Kelvin's advocacy of proper industrial standards, based on this case, was to give Britain a great advantage over its industrial competitors in later years.

His dogged persistence in the campaign for proper units and standards is exemplified by a comical excerpt from a verbatim report on Electrical Standards Committee of the Board of Trade (1891), in Strutt's biography of his father, Lord Rayleigh [4].

> *The Chairman*: I am bound to tell the Committee that I did speak to Sir Michael Higgs Beach [President of the Board of Trade], about this particular definition, and that it certainly frightened him a great deal.
> *Sir William Thomson*: A page from the theory of averages would frighten him still more.

Kelvin was not prepared to be obstructed by bureaucratic blockheads.

It is sad to reflect that even this huge economic benefit to the British economy did not, of itself, provide the motivation for his peerage, which was conferred only when he strayed into active politics.

3. Foam

By 1887, he was ideally qualified to design a material ether, and his musings in bed were quickly turned into a very strange model of an ether that consisted of a liquid foam without any enclosed gas.

To see that this should have the required mechanical properties was an extraordinary feat of imagination, and we have no idea where it came from. Maxwell wrote a poem about "empty bubbles" rising out of the unconscious during sleep, which is curiously apt to describe Kelvin's revelation, but what more rational sources it may have had are obscure. It was a totally new and utterly brilliant idea, however out of tune with a time when the concept of ether was becoming obsolete.

The mathematical result that he conjectured has turned out to be incorrect. It was supposed to answer the question: what foam structure, made up of bubbles of equal size, has the lowest surface area, and hence energy? Although eventually surpassed, the Kelvin foam structure remains of interest today – truly a triumph of failure.

From the very outset his colleagues were not impressed [1] ("utterly frothy", "a pure waste of time" etc.) [5]. They were particularly inspired by the experiment of Hertz, the youngest of the Maxwellians, and its Maxwellian interpretation by Fitzgerald. Many scathing comments were made (mostly in private), such as that of Rayleigh to Schuster [4]:

> "Sir W. is full of a *froth* theory of the ether! This will lend itself to sarcasm even better than the jelly theory".

The reference is to an earlier notion of Stokes, which Kelvin had also found interesting [6].

Even compendious histories of the ether pay little attention to Kelvin's foam. In his long list of 650 or so papers, his proposal [1] of an ideal foam structure, intended for the ether, has been considered relatively insignificant by biographers. Nevertheless it would rank high in terms of the fashionable citation analysis of today. Mathematicians, physicists, materials scientists and biologists have all found it stimulating, when they consider foam or other cellular structures.

So it is still worthy of a close reading, and accordingly a critique of it has recently been published [5].

Perhaps because the paper was composed in some haste and probably not subjected to review before publication (since Kelvin was the Editor), it is hardly a model of clarity. It recounts the author's original thought processes rather than some reconstructed logic later formulated in tranquillity, in the interests of transparency. Just what we tell graduate students not to do?

That first paper, obviously intended to lay the foundation for much that was to come (but never did) was entitled:

On the Division of Space with Minimum Partitional Area

It was confined to the problem of the specification of the foam structure of equal bubbles that has minimal surface area: the ether is not mentioned! The question addressed, and encapsulated in the title, is today called the Kelvin Problem [6].

Kelvin's instincts and logic led him to what he called the *tetrakaidecahedron,* as the shape of the cells packed together in the foam. Today's physicists would relate it to the Wigner-Seitz cell of the bcc structure, although it is not precisely the same, in that some of its faces are subtly curved. Kelvin's mathematical treatment of this difficulty is highly impressive, involving judicious approximation and numerical calculation, which he got right.

In general this kind of problem is impossibly difficult without the aid of computers. Only in the last decade or so has it become routinely possible to compute such surfaces accurately.

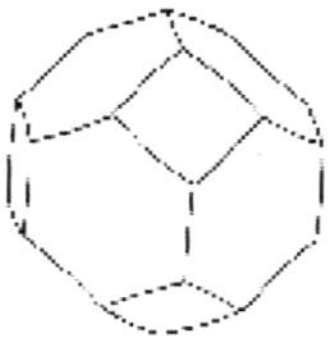

Figure 1
The Kelvin Cell, or tetrakaidecahedron, consists of six flat quadrilateral and eight hexagonal faces, which have a subtle curvature. It can be packed to fill all space, in a body-centred cubic arrangement.

4. Later developments

The unresolved Kelvin Problem [7] has lingered in the minds of mathematicians: *was his suggested structure really the best?*

If so, one might hope to see it in nature, but even individual Kelvin cells proved to be very rare. Was this because of the trapping of real macroscopic structures in haphazard disordered configurations, or the variation in cell sizes, or was it because Kelvin's structure was not optimal after all? Opinion remained divided until the 1990's.

In 1994, Weaire and Phelan [8] published the first counter-example to the conjecture that the Kelvin structure has lowest surface area.

Nevertheless, the Kelvin structure lives on. Only recently we have found out how to make it. Ironically, we cannot yet make the superior Weaire-Phelan foam in the laboratory. But it does loom large in a somewhat artificial manifestation in 2008.

5. Bubbles in Beijing

The Water Cube, the aquatic centre of the Olympic Games, consists of a gigantic steel framework, clad in plastic. Its beams represent the Plateau borders of the Weaire-Phelan structure. The designers started with the Kelvin structure, in a search for something emblematic of water, but eventually turned to the more irregular (but nevertheless periodic)WP structure, as conforming better to today's aesthetic [9].

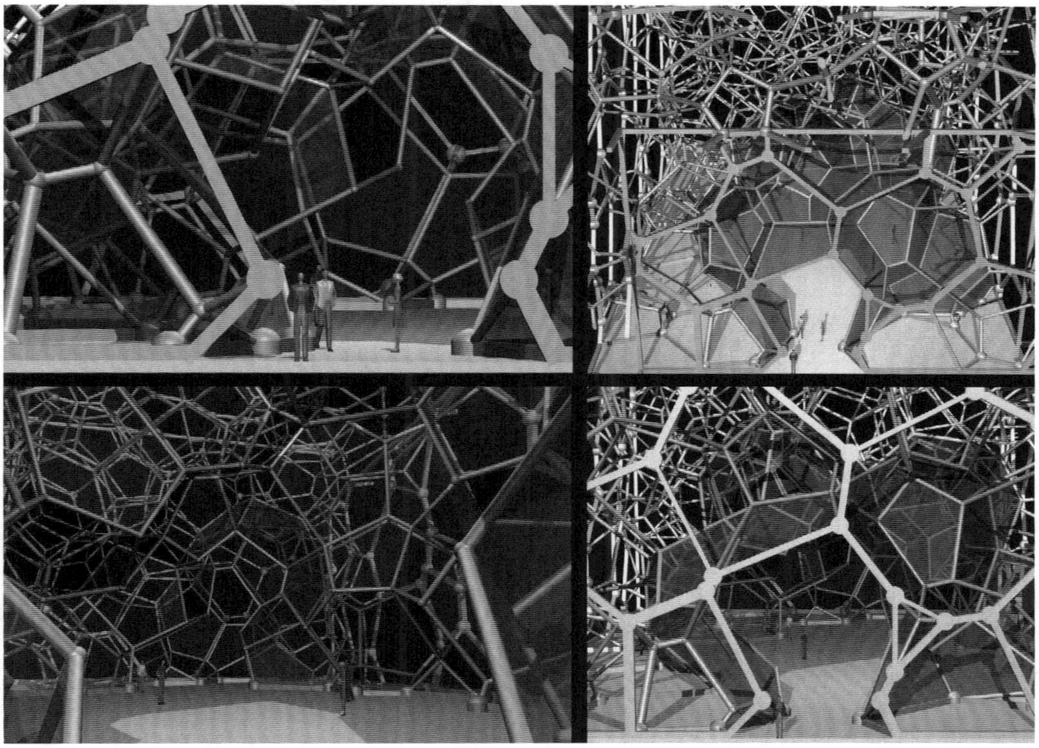

Figure 2. Part of the original design for the interior of the Water Cube, based on the Weaire-Phelan Structure. This is a more complex structure than that of Kelvin, with two kinds of cells. Pentagonal faces predominate.

It will be admired by many millions. For the physicist, it may be a monument to Kelvin's restless curiosity, expressed in the Kelvin Problem, which as a problem in rigorous geometry still remains open for future generations of geometers.

Acknowledgements

Research support is acknowledged: ESA/ESTEC, Science Foundation Ireland and a CNRS visiting appointment in 2007.

Bruce Hunt is thanked for his authoritative advice.

Journal of Physics: Conference Series **158** (2009) 012005 doi:10.1088/1742-6596/158/1/012005

References

[1] Thomson W 1887 *Philosophical Magazine* **24**, 503
[2] Weaire D 2008 Kelvin and Fitzgerald: great Irish physicists *Kelvin: Life, Labours and Legacy* ed Flood R, McCartney M and Whitaker A pp 86-93
[3] Hunt B 1991, *The Maxwellians* (Ithaca: Cornell University Press)
[4] Strutt R J 1924 *John William Strutt, Third Baron Rayleigh* (London: Edward Arnold)
[5] Weaire D 2008 Kelvin's foam structure: a commentary *Philosophical Magazine Letters* **88** 91-102
[6] Wilson D B 1990 *The Correspondence between Sir George Gabriel Stokes and Sir William Thomson, Baron Kelvin of Largs* (Cambridge: Cambridge University Press)
[7] Weaire D (ed) 1996 *The Kelvin Problem* (London: Taylor and Francis)
[8] Weaire D and Phelan R 1994 The structure of monodisperse foam *Philosophical Magazine Letters* **70** 345-50
[9] Aste T and Weaire D 2008, *The Pursuit of Perfect Packing*, (London: Taylor and Francis) (2nd Edition)

Journal of Physics: Conference Series **158** (2009) 012006 doi:10.1088/1742-6596/158/1/012006

Is Kelvin representative of an "Irish tradition"?

Andrew Whitaker
Department of Physics, Queen's University, Belfast BT7 1NN,
Northern Ireland

E-mail: a.whitaker@qub.ac.uk

Abstract. Kelvin interacted with many physicists and mathematicians with Irish connections, and these interactions are sketched. A number of suggestions of Irish or Irish-Scottish scientific traditions or networks are reviewed, but it is argued that the large number of interactions may be better explained just by the high quality of physics and mathematics in Ireland and Scotland in the nineteenth century. Kelvin's disapproval of science as carried out in Cambridge and London is discussed, and an attack by some of his followers on science at Oxford is also described. Finally a number of examples of Kelvin's most important work are outlined and some general features are pointed out; these include the willingness to tackle practical and engineering problems, and the ability to attempt to construct major syntheses, where appropriate taking note of religious themes, and drawing conclusions, positive or negative, on religious matters. These features are used to construct a highly tentative characterisation of a grouping of Irish and Scottish scientists centred largely round Kelvin.

1. Kelvin and nineteenth century Irish physicists and mathematicians

It has often been remarked that Kelvin had a very high regard for Irish mathematics and physics. As an example, when he became Editor of the *Cambridge Mathematical Journal* in 1845, he changed its name to the *Cambridge and Dublin Mathematical Journal* in order to bring in papers, as he said, from all over Ireland, not just from Dublin [1, 2]. (Refs. [1] and [2] are standard excellent biographies of Kelvin that may be consulted with regard to most aspects of his life and work.)

Kelvin also learned from, or collaborated or interacted with, many mathematicians or physicists who were Irish or who had Irish connections. To establish the point, a short list will follow. Most of those listed have brief biographies in either or both of Refs. [3] and [4].

James MacCullagh (1809-47) (figure 1) [5] was an important theoretical physicist from Trinity College Dublin (TCD). His main work consisted of study of the passage of light through solids. We may speak of this as an early contribution to study of the ether, the hypothetical medium through which it was supposed that light must travel. As we shall see, many of the most celebrated Irish scientists of this century concentrated their efforts on study of the ether, and Kelvin himself was heavily influenced by MacCullagh. Sadly MacCullagh's scientific and political ambitions were not realised, and he was to commit suicide at a comparatively early age.

Today the work of George Boole (1815-64) (figure 2) [6] is known by anyone who uses either statistics or a computer, in the form of Boolean logic. Boole, an Englishman, had no qualifications in mathematics, and much of his early work was performed as an amateur. Kelvin was among those who helped him to obtain his first position as a mathematician, when, at the foundation of the Queen's Colleges in Ireland, he became the first Professor of Mathematics at Queen's College Cork in 1849.

Journal of Physics: Conference Series **158** (2009) 012006 doi:10.1088/1742-6596/158/1/012006

He too died early, catching pneumonia after becoming soaked while insisting on walking through heavy rain to give a scheduled lecture.

Figure 1
James MacCullagh

Figure 2
George Boole

Figure 3
William Rowan Hamilton

Figure 4
Peter Guthrie Tait

William Rowan Hamilton (1805-65) of TCD (figure 3) [7-9] was, of course, one of the most creative mathematicians of all time. Today his most celebrated achievements would probably be seen as his important work in optics, including the discovery of conical refraction, and the so-called Hamiltonian mechanics, a hundred years later an indispensable part of quantum theory, with its ubiquitous Hamiltonian function. Kelvin, of course, fully appreciated this early work. However Hamilton himself was most proud of his discovery of the abstract algebra of quaternions. Although most of this work was published under Kelvin's editorship of *The Cambridge and Dublin Mathematical Journal*, Kelvin came to consider that "Quaternions came from Hamilton after his really good work had been done, and though beautifully ingenious, have been an unmixed evil to those who have touched them in any way, including Clerk Maxwell" [10]. One such, in Kelvin's opinion, was Peter (PG) Tait (1831-1901) [11], in every other way one of Kelvin's closest friends and collaborators.

Tait (figure 4) was a Scotsman and a Cambridge Senior Wrangler [top performer in the Mathematical Tripos], who was Professor of Mathematics at Queen's College Belfast (QCB) from 1854-60 before moving to become Professor of Natural Philosophy (Physics) at the University of Edinburgh, a Chair that he occupied until his death. His move from Mathematics to Physics paralleled that of Kelvin. In obtaining the Edinburgh position, he defeated James Clerk Maxwell, an obviously superior physicist, presumably because of Tait's superior teaching ability. The Edinburgh *Courant* commented that: "[I]n Professor Maxwell the curators would have had the opportunity of associating with the University one who is already acknowledged to be one of the remarkable men known to the scientific world...But there is another power which is desirable in a professor of a University with a system like ours, ... the power of oral exposition.... We little doubt that it was the deficiency of this power in Professor Maxwell principally that made the curators prefer Mr Tait [11 pp 16-17]."

It might be said that Tait acted as Kelvin's "bulldog" in his arguments with evolutionists and geologists over the age of the Earth [12]. With Tait's support, Kelvin estimated the total amount of energy that would have been available over the life of the Universe. From this comparatively small amount of energy, they deduced that the lifetime of the Earth must have been far shorter than required by the contemporary theories of evolution and geology. While Kelvin's arguments were broadly correct according to the then state of knowledge, it is usually argued that the discovery of the additional energy supply of radioactivity towards the very end of both his and Tait's lives was to render their age limits far too low. A recent paper [13], though, suggests that the real difficulty was that Kelvin had assumed that the Earth had a solid rather than a liquid interior, and so his model did not include convection. The argument continues [14, 15]. Though Kelvin himself was only a mild opponent of evolution, his arguments with Huxley, Darwin's own "bulldog" (figures 5-7), mainly over the timescales involved for evolution, have damaged his posthumous reputation rather severely.

Figure 5
Thomas Henry Huxley

Figure 6
(from left) Michael Faraday, Huxley, Charles Wheatstone, David Brewster, John Tyndall

Figure 7
Charles Darwin

A more important interaction between Tait and Kelvin was the writing between 1861 and 1866 of the famous *Treatise on Natural Philosophy* [16], always known, after Thomson and Tait, as *T and T'*. This book followed the discovery of the Laws of Thermodynamics by Rudolf Clausius (1822-88) (figure 8), Kelvin, and William John MacQuorn Rankine (1820-72) (figure 9), and in particular the understanding obtained by Kelvin, with the assistance of his brother, James Thomson, of the significance of these Laws in the entire history of the Universe, and of the centrality of the concept of energy throughout this history. *T and T'* was the first book to be based around the paradigm of energy rather than Newtonian force, and so initiated the approach to physics taken for granted by any student of the subject today. As has already been stated, Tait was a great enthusiast for quaternions. Kelvin's rejoinder was that he would be happy for their inclusion in the book if Tait were able to show that they had ever been of any use, but Tait was unable to do so [17]!

Figure 8
Rudolf Clausius

Figure 9
William John MacQuorn Rankine

Figure 10
Thomas Andrews [Queen's University Belfast]

Figure 11
Statue of Henry Cooke in Belfast

Figure 12
Pooley Henry [Queen's University Belfast]

Another scientist with Belfast connections well-known to Kelvin was Thomas Andrews (1813-85) (figure 10) [18], who had been a pupil of Kelvin's father, James Thomson Senior, at the Belfast

Academical Institution. With the foundation of the Queen's Colleges in 1845, it seemed highly likely that James, whose prominence by that time at Glasgow University was well-known, would be asked to become the first Principal of QCB. However, by mid-century, the political liberalism of Belfast in the first third of the century had largely disappeared. Henry Cooke (1788-1868) (figure 11), the popular preacher and leader of the Presbyterians in Belfast, had succeeded in breaking the 1798 alliance of Catholic and dissenter, and established the broad division between Catholic on the one hand and all aspects of Protestantism on the other.

Cooke was keen to become President of QCB himself, but, as a minimum requirement, he made it clear that, if any appointments were made to QCB of whom the Presbyterian Church did not approve, no trainee for the ministry would be allowed to attend the college. As it happened, the liberal Thomson and the conservative Cooke were both ruled out, and the position went to Dr Pooley Henry (figure 12), neither an academic nor an educationalist, but as an administrator a safe pair of hands. Thomson was offered the position of Vice-President at a little over half the salary of the President, a position he naturally turned down [19, 20].

The positive result of this debacle was that the position of first, and, as it turned out, only Vice-Principal of QCB went to Andrews, an excellent scientist, who was to carry out the famous Andrews' experiment, which clarified the difference between a vapour, which may be liquefied by pressure alone, and a gas which may not. This was to be crucial over the future decades as liquefaction of the various gases was steadily achieved.

Figure 13
George Stokes

Figure 14
John Tyndall

Figure 15
Michael Faraday

Figure 16
James Joule

Figure 17
Julius Meyer

George Gabriel Stokes (1819-1903) (figures 13 and 71) [21-23] came from County Sligo, and was Lucasian Professor at Cambridge from 1849, as well as being Secretary of the Royal Society for many years and later President. He made discoveries in many areas of physics, being particularly concerned with the properties of the ether, and there are several well-known Stokes' Laws and Theorems. Despite their geographical separation, he was the scientist closest to Kelvin, and they exchanged innumerable letters [24-25]. Stokes, an Anglican, was an extremely devout man, and was concerned with the links between science and religion.

On the other hand, John Tyndall (1820-93) (figures 6 and 14) [26-29] from Carlow could be described as Kelvin's greatest scientific opponent, if not enemy. With the support of Michael Faraday (figures 6 and 15), Tyndall became Professor of Natural Philosophy at the Royal Institution and made contributions to many areas of physics, the most important perhaps being the physics of glaciers, as well as being a leading communicator and lecturer. He was a follower of Huxley and a fellow agnostic, and his famous "Belfast address" to the British Association for the Advancement of Science (BAAS) in Belfast in 1874 was a trenchant appeal for materialism, and a demand that any religious input to science should be eliminated. It was clear that his approach to life was entirely opposed to that of Kelvin, but their animosity seems to have gone deeper. It may be that the self-made Tyndall resented the fact that Kelvin had, in comparative terms, rather a privileged upbringing and obtained a prestigious position at a young age. Tyndall's role in diminishing the claims of Kelvin's protégé, James Joule (figure 16) in favour of those of Julius Mayer (figure 17) is discussed in section 3. On

Tyndall, Kelvin undoubtedly shared Tait's views; describing William Grove (1811-96), a well-known physicist of the time, as "woefully loose and unscientific... vague and cloudy", he still considers him "not [as bad as] a Tyndall, as of course you know, since the latter is unique" [29, p 176].

Figure 18
Samuel Haughton
© Trinity College Dublin

Figure 19
Joseph David Everett
[Queen's University Belfast]

Another TCD scientist with definite academic links to Kelvin was Samuel Haughton (1821-97) (figure 18) [30], Professor of Geology from 1851, and later, after gaining a medical qualification, Registrar in the School of Medicine, where he was placed by the Provost and College authorities to reform the rather chaotic administration of the School. As a geologist, he was a strong supporter of Kelvin in the controversy over the ages of the Earth and the Sun, and, like Kelvin, his reputation was to suffer when it was shown that their conclusion, if not necessarily their argument, was incorrect. Rather unfairly Haughton may be best known for the so-called "Haughton drop", the formula based on the weight of a person to be hanged, which ensured death would be instantaneous due to a broken neck, rather than over a considerable period of strangulation.

John Everett (1831-1904) (figure 19) [31] was originally from England, but studied under Kelvin in Glasgow. He was undoubtedly Kelvin's best student in the 1850s [32], and probably one of the two or three best in the whole of Kelvin's 53 years as Professor in Glasgow. Everett subsequently became Professor of Natural Philosophy at QCB from 1867-97, and during these years, as well as writing influential textbooks, he became prominent in the task of establishing a set of electrical and other units.

In this project he was following in the footsteps of Kelvin. From the time of his earliest considerations of the Atlantic cable, Kelvin had recognised the importance of accurate determination of the conductance of the copper and the capacitance of the cable as provided by the manufacturers. However this was stymied by the fact there were no national or international standards for electrical quantities. Thus Kelvin was to spend a considerable amount of time taking part in and chairing International Commissions, which not only had the task of determining the most practical means of defining the various units, but also the sensitive one of allocating the names of prominent scientists to the various quantities. Everett took the process further, and his most important role was as Secretary to the British Association Committee for the Selection and Nomenclature of Dynamic and Electrical Units. In 1873 he presented a report recommending the cgs or centimetre-gram-second system of units, and introducing the names of *dyne* and *erg* for the fundamental units of force and energy in this system [33]. Today, when we use the mks or metre-kilogram-second system, our units are the *Newton*, which is equal to 10^5 dynes, and the *Joule*, which is equal to 10^7 ergs.

George Francis Fitzgerald (1851-1901) (figure 20) [34-36] was yet another famous TCD scientist. He is probably most famous for the Fitzgerald(-Lorentz) contraction, the attempted solution to the dilemma of the Michelson-Morley experiment, the discovery that it appeared to be impossible to demonstrate motion through the ether. For Fitzgerald, the suggestion, though certainly brilliant, was made rather casually. Formally the contraction reappears in Einstein's theory of special relativity, though here the interpretation is rather different; it is an *apparent* contraction, rather than the *real* contraction imagined by Lorentz and Fitzgerald. In Einstein's theory, the impossibility of demonstrating motion through the ether is explained by the postulate that the ether does not actually exist.

Fitzgerald, though, did much more significant work. Following James Clerk Maxwell's production of his epoch-making theory of electromagnetism (figure 21), Fitzgerald became the unofficial spokesman for the so-called Maxwellians [37-38] whose self-imposed task was to elucidate and

Figure 20
George Fitzgerald

Figure 21
James Clerk Maxwell

Figure 22
Oliver Lodge

Figure 23
Oliver Heaviside

Figure 24
Heinrich Hertz

promote all the various aspects of Maxwell's theory. Apart from Fitzgerald himself, this group included Oliver Lodge (1851-1940) (figure 22), Oliver Heaviside (1850-1925) (figure 23), Heinrich Hertz (1857-94) (figure 24), who was the first to detect the electromagnetic waves predicted by Maxwell, and Joseph Larmor, who will appear later in this list. Earlier in his career Fitzgerald had been a great admirer of Kelvin, but Kelvin's opposition to Maxwell's theory inevitably meant that relations between the two became at least temporarily somewhat strained. However, when Kelvin had occupied the Glasgow chair for 50 years in 1896, Fitzgerald contributed the main biographical article to the celebratory volume, presumably signifying his willingness to accept the positive contributions of the older man and ignore those of which he disapproved [39].

Figure 25
John Perry

Figure 26
William Ayrton

Figure 27
The Ayrton and Perry Trike of 1881, believed to be the first road going electric vehicle [Bluebird]

John Perry (1850-1920) (figure 25) [40, 41] was an extremely interesting character who cropped up in Kelvin's life at a number of significant points. He was born in Garvagh in County Londonderry, and though he left school at the age of 14 to take up an apprenticeship, he gained scholarships to study at QCB, working under James Thomson and obtaining his MA. As mathematics master and physics lecturer at Clifton College in Bristol from 1870-84, he set up the first physics laboratory and the second mechanics workshop in an English school. He subsequently worked as an assistant to Kelvin for a year, and then, with William Ayrton (figure 26), took Kelvin's influence to Japan as joint

Professors of Engineering [42]. An important contribution to physics of the pair during this period was an accurate determination in 1878 of the speed of light using telegraphy; this value was considerably closer to that predicted by Maxwell than those of the previous best measurements, giving enhanced experimental backing to his theory [29].

From the 1880s, Perry and Ayrton were both back in England as Professors at Finsbury Technical College, later to become part of Imperial College London. Perry was Kelvin's main critic from the physics world in the controversy over the age of the Earth [12]. He actually considered Kelvin's arguments to be correct in principle, but suggested that he had used detailed assumptions which were somewhat arbitrary, and which affected his results quite significantly. Perry preferred the liquid to the solid interior of the earth [13-15].

During this period, Perry and Ayrton published many scientific papers, and were also responsible for a vast number of inventions, for example the first block system for electric railways, the first electric tricycle (figure 27) and many electrical measuring instruments. Perry was a great champion of technical education, giving classes for working men, and also of mathematics for engineers and scientists, and he published many textbooks in this area. He also made a very great improvement to the slide rule (the so-called System Perry) [43], and stressed the importance of graph paper, at the time extremely expensive; Perry was able to achieve an enormous reduction in cost. Perry and Ayrton will appear again shortly in this account.

Figure 28
Joseph Larmor

Figure 29
Paul Dirac

Figure 30
John Townsend

Another very important figure, much younger that Kelvin, who interacted with him very significantly, was Joseph Larmor (1857-1942) (figure 28) [44-46]. He was a native of County Antrim, and he studied first at QCB, before moving to Cambridge where he became Senior Wrangler in 1880. He was Professor of Natural Philosophy at Queen's College Galway (QCG) from 1880 to 1885, subsequently returning to Cambridge as a lecturer before succeeding Stokes as Lucasian Professor in 1903. He himself was succeeded in this position by PAM Dirac (figure 29) in 1932.

Larmor's important researches were in electromagnetism and the ether, and he had many discussions with Kelvin in this area of physics. It is now recognised that he was the first to write down the so-called Lorentz transformation equations, which are of crucial importance in special relativity. Larmor is best known for the his discussion of the radiation from a moving charged particle, while in the theory of magnetic resonance, the Larmor precession and its associated Larmor frequency are absolutely central. Larmor edited the collected works of both Stokes and Kelvin.

Last to be mentioned in this brief list is John Townsend (1868-1957) (figure 30) [47], who came from Galway, and studied at TCD under Fitzgerald and worked with JJ Thomson in Cambridge on ion physics before becoming the first Wykeham Professor of Physics at Oxford in 1900. By this time

Kelvin was in his mid-70s, but he wrote references for Townsend, and, once the latter had been appointed to the Oxford position, took considerable pains to assist in him obtaining a Royal Society grant for equipment [48].

2. Was there an 'Irish tradition'?

It is natural to speculate whether Kelvin's interactions, some negative but the great majority positive, with scientists from Ireland show that he was part of an Irish scientific and mathematical tradition. This speculation may be followed up in several very different directions.

First we may note the argument of Ivor Grattan-Guinness [49, also 44] that there is a clear line in the study of the ether from the physics of MacCullough and the mathematics of Hamilton to the radical ideas of Fitzgerald and thence of Larmor. Haughton also carried out some admittedly less important work along the same lines, and it is noticeable that, of these five, only Larmor was not from TCD. Kelvin and Stokes also spent large portions of their careers studying models of the ether, so it is clear that Irish mathematicians and scientists were important contributors to such work throughout the nineteenth century.

In *Prometheus' Fire*, Norman McMillan [28] argues that the idea of an Irish tradition is insufficient to explain "the importance and success of the Irish in British science and engineering", and suggests instead the idea of an *Irish network*, not in any way as formalised as the X-club of Huxley and colleagues, but nevertheless a group of people with quite strong social cohesion, prepared to exchange advice and opinions, and also to help each other into positions of importance and influence, and provide assistance in gaining fellowships and honours.

Figure 31
Edward Sabine
[© National Portrait Gallery, London]

Figure 32
Thomas Grubb
[UCSC Special Collections: Lick Observatory Records]

Figure 33
Howard Grubb
[UCSC Special Collections: Lick Observatory Records]

Figure 34
William Parsons
3rd Earl of Rosse

McMillan includes in his discussion Edward Sabine (1788-1883) (figure 31), a military engineer and explorer originally from Dublin, who was President of the Royal Society (RS) from 1861-71, but had been Secretary, Foreign Secretary and then Treasurer for a total of 19 years before that. He had also been Secretary to the BAAS from 1839-59 except for the year of 1852 when he acted as President. McMillan mentions that Sabine assisted Stokes and Tyndall to become FRS in 1851 and 1852 respectively; he hoped that they would aid him in the continuing process of reforming the society. Sabine also actively worked for Kelvin's election in 1851; in this case his particular aim seems to have been to achieve theoretical support for Sabine's own labours in terrestrial magnetism [50].

McMillan describes in more detail Stokes' work as Secretary of the RS from 1854-85. (He was later President from 1885-90.) McMillan discusses informatively the support he gave to Tyndall, despite the fact that the two had drastically divergent views on religious, and to an extent, social matters; to Howard Grubb (1844-1931) (figure 33) [51], most famous, with his father, Thomas (figure 32), as a designer and manufacturer of telescopes, and whose optical company received a vast amount of technical advice from Stokes; to the Third Earl of Rosse (1800-67) (figures 34 and 71) [52], responsible for the construction at Birr of the largest telescope in the world at that time; to Robert Mallett (1810-81) (figure 35), founder of the science of seismology and vulcanology; to John Casey (1820-91)(figure 36) [53], a poverty-stricken schoolteacher, who Stokes helped to eminence as the most famous mathematician who has been associated with the Catholic University of Ireland and its successor institution, University College Dublin (UCD); to Fitzgerald, as he commenced his analysis of Maxwell's work; and to several others with Irish connections, including Haughton.

Figure 35
Robert Mallett
[Engineers Ireland]

Figure 36
John Casey
[London Mathematical Society]

Figure 37
George Johnstone Stoney
[reproduction by kind permission of the Royal Dublin Society]

Figure 38
Humphrey Lloyd
[© The Royal Society]

McMillan points out that Stokes, who could select from papers submitted to the Royal Society those he wished to review, in fact reviewed papers written by 22 Irish authors, which represented a statistically higher proportion than for non-Irish authors. Of his help to scientists, McMillan says that: "It seems a historically determined fact that he reserved most of his help for those from his homeland." Indeed McMillan says that archives of other Irish scientists, including Sabine, Tyndall. Fitzgerald and George Johnstone Stoney (figure 37), show the same Irish "connectivity", and he calls for a careful study of "the many successes in the government of British science and engineering through the active co-operation of the Irish in furthering their own careers and their own specific national interests". (Stoney [54], who was Professor of Natural Philosophy at QCG from 1852 to 1857, and then Secretary to the Queen's University administration in Dublin until 1882, was also a prolific physicist, best known today for having named the electron.)

An interesting test of these ideas is whether Kelvin received particular support from Irish people when he applied for the Glasgow chair in 1846. Smith and Wise [1, p 277] comment that, at their 1844 meeting: "No doubt Sabine represented to Thomson the possibility of support in high places for the forthcoming campaign for the Glasgow chair." Help behind the scenes may have been forthcoming, but Sabine is not on the list of those providing testimonials given by Thompson [1, pp 167-8], which includes Hamilton, Stokes, Boole and Humphrey Lloyd (1800-81) (figure 38) [55], thus only 4 with strong Irish connections out of around 30 on the list. (Lloyd was an experimental physicist, later Provost of TCD; today best known for the so-called Lloyd's mirror, a clear demonstration of the wave nature of light, but probably more important for his setting up of Magnetic Observatories at TCD and throughout the world.)

McMillan's ideas on the network were originally put forward specifically for Irish astronomers [56]. Nicholas Whyte [57], however, questions the use of the term *network*. Whyte's own model of a network is the so-called Cambridge Network (figures 39-44), which was active in the second quarter of the nineteenth century. It included scientists such as Charles Babbage, John Herschel, George Peacock, George Biddell Airy, William Whewell and Adam Sedgwick, and its self-imposed task was the reform of British science, which involved modernising British mathematics and in particular its teaching at Cambridge, attempting to obtain control of the Royal Society, and stimulating the BAAS. To this end they worked together and provided mutual support, and, to an extent, patronage. Whyte does not believe that these elements were present in the Irish case.

Figure 39
Charles Babbage

Figure 40
John Herschel

Figure 41
George Peacock

Figure 42
George Biddell Airy

Figure 43
William Whewell

Figure 44
Adam Sedgwick

However he supports the idea of an *Irish astronomical tradition*. He notes the foundation of the Dunsink Observatory near Dublin in 1775, and the Armagh Observatory in 1792. As well as several of the characters we have already met – Hamilton, Rosse and the Grubbs, Whyte includes in the tradition Thomas Romney Robinson (1793-1882) (figure 45), astronomer in Armagh; Edward Joshua Cooper (1793-1863) (figure 46), who built up another flourishing private observatory; Sir Robert Ball (1840-1913) (figure 47), astronomer at Dunsink, and later Professor of Astronomy at TCD and Cambridge, and while at TCD, Astronomer Royal for Ireland; Agnes Clerke (1842-1907) (figure 48), who wrote a number of extremely popular histories of astronomy and cosmology; and many others. William Wilson (1851-1908) (figure 49) was also an important Irish astronomer with a series of telescopes at his family estate of Daramona.

Figure 45
Thomas Romney Robinson
[The Irish Picture Library/Birr Castle Collection]

Figure 46
Edward Joshua Cooper

Figure 47
Robert Ball

Figure 48
Agnes Mary Clerke

Figure 49
William Edward Wilson
[University of London Observatory: University College London]

Journal of Physics: Conference Series **158** (2009) 012006 doi:10.1088/1742-6596/158/1/012006

Whyte also points out another feature tending to create communication and support among Irish astronomers and indeed the wider scientific community – the fact that there were many family and other personal relationships. Stokes married one of Robinson's daughters. Robinson himself was the uncle of WH Rambaut, assistant at Birr and Armagh in the mid nineteenth century, who was himself the uncle of AA Rambaut, director at Dunsink at the end of the century. Ball's first job was as tutor to Rosse's children, who included Charles Parsons (figure 50), later a distinguished engineer, who was to buy the company founded by Thomas Grubb. GJ Stoney had a brother, Bindon Blood Stoney (1828-1909) (figure 51), who was famous as the engineer responsible for Dublin's harbour, and he was also the uncle of George Fitzgerald and Maurice Fitzgerald, George's brother, who was Professor of Engineering at QCB. And Whyte gives many more examples. Incidentally McCartney [47] notes that, at Cambridge, Ball made a point of giving hospitality and support to those from Ireland, including Townsend.

Figure 50
Charles Parsons

Figure 51
Bindon Blood Stoney [Engineers Ireland]

Before analysing these ideas, we note that it might seem entirely natural to extend the tentative idea of a tradition to include Scotland as well as Ireland. This would recognise that Kelvin had his whole professorial career in Glasgow, Tait was of course very much a Scotsman, and several others of those mentioned, such as Everett and Perry, worked with Kelvin in Scotland. Also it particularly recognises the close relationship of Kelvin with Maxwell, who, could be regarded as Kelvin's 'protégé' [58]. In their lively and often jocular communications it was as natural for [JC] Maxwell to be dp/dt, because of the well-known physics equation written in standard notation as $dp/dt = jcm$, as for Kelvin and Tait to be *T* and *T'*.

Crosbie Smith [29] takes this further in his book *The Science of Energy* by introducing an "identifiable, though informal, network of scientific practitioners" which he calls the *North British group*, who worked strenuously in the mid nineteenth century to produce a theory of energy and to make practical use of that theory, but worked equally strenuously to establish practically total priority for the members of their group, and so to establish their own reputations as among the most important scientists of all time. Smith picks out, as locations for the work and promotion of the network, Scottish Universities, marine engineering works, and local and national scientific societies, in particular the BAAS. He gives the main names involved as Kelvin, Maxwell, Joule and Rankine, with Tait, Fleeming Jenkin (figure 52) and James Thomson (figure 53) standing somewhat lower than these.

Of the newer characters, James Joule (1918-89) (figure 16), an independent scientist from Manchester, is generally recognised as establishing the mutual interchangeability of heat and work, and thus, particularly in the view of the North British scientists, providing the First Law of

Journal of Physics: Conference Series **158** (2009) 012006 doi:10.1088/1742-6596/158/1/012006

Thermodynamics. Fleeming Jenkin (1833-85) was born in England but went to school in Edinburgh, finally graduating at Genoa. He was working as an engineer in England when he became associated with Kelvin, assisting him on telegraph cables and many of his inventions. Jenkin was actually a man of considerable erudition, writing on a wide range of scientific and conceptual issues, and he became Professor of Engineering at Edinburgh in 1868.

Rankine (Fig 9) was originally from Edinburgh and worked principally as a railway engineer before becoming interested in the late 1840s in the study of heat, work and engines. His model of *molecular vortices* provided a clear example of the conversion of heat to work. (The reverse process is less problematic, and may be said to have been demonstrated in the much earlier experiments using friction of Sir Humphry Davy (figure 54) and Count Rumford (figure 55), even though the significance of these was not widely accepted at the time.) Rankine's work strongly influenced that of Kelvin, and, as already stated, Rankine may be regarded as one of the three founders of thermodynamics, along with Kelvin and Clausius (of whom, see later), although Rankine's work, being based on models, could be regarded as less profound than that of Kelvin or Clausius. In the 1850s, Rankine helped to establish the practical credentials of thermodynamics by designing actual engines, as another engineer, James Thomson, Kelvin's brother, had done earlier, and Rankine became Professor of Civil Engineering at Glasgow in 1855. Following Rankine's death, James Thomson would be his successor.

Figure 52
Fleeming Jenkin

Figure 53
James Thomson

Figure 54
Humphrey Davy

Figure 55
Benjamin Thompson
Count Rumford

The term *North British* is usually used as little more than a synonym for *Scottish*, rather than referring, for example, to the northern parts of England and Ireland as well. (If it were to reach as far south as Manchester and Joule, it would have to cover a very large fraction of the British Isles!) When Smith (p 313) wishes to include Ireland he specifically refers to *West* as well as *North Britain* [59]. Thus, in this excellent book, an extremely important aspect of Kelvin's work appears to be divorced from Irish links [60]. We shall return to the question on Kelvin – (simplistically) Irish or Scottish? – later.

We have looked at suggestions for a variety of different types of group and alliance, and doubtless all contain much truth; certainly each may be used as the basis for an informative account of the history involved. (Another interesting paper considering the question of an Irish tradition specifically in mathematics has been written by Raymond Flood [61].) Yet one may wonder if at least part of the reason for Kelvin's interaction with many Irish and Scottish mathematicians and physicists is just that many of the best mathematicians and physicists in his lifetime were Irish and Scottish!

Figure 56
Nicholas Callan
[Maynooth College]

Figure 57
John Joly
[Reproduction by kind permission of the Royal Dublin Society]

Figure 58
Thomas Preston

Figure 59
Frederick Trouton

Of the Irish contingent, one might supplement our original list and those given since by the names of, for example, Nicholas Callan (1799-1864) (figure 56) [62], Professor of Natural Philosophy at Maynooth College, where Catholic clergy were trained, and known as the inventor of the induction coil; John Joly (1857-1933) (figure 57) [63], another great TCD figure, famous for many inventions including the steam calorimeter, who also made advances in geology and geophysics; Thomas Preston (1860-1900) (figure 58) [64], Professor of Experimental Physics at UCD, discoverer of the anomalous Zeeman effect and author of two famous physics textbooks; and Frederick Trouton (1863-1922) (figure 59) [65], originally from TCD, but from 1902 Professor of Physics at University College London, best-known today for "Trouton's law" relating boiling points and latent heats, but whose most important achievements were his experimental studies of the applications of Maxwell's Laws. More names could be added, as is true also for Scotland.

Indeed Herries Davies [66] comments that: "Quite simply, many of Ireland's scientists in the period 1780 to 1880 were among the world's leading scientific figures." It is also of interest that in the earliest days of the BAAS [67-8], Ireland and of course, Scotland were prominent. After the first meeting of 1831 in York, and meetings in the following years in Oxford and Cambridge, the 1834 and 1835 meetings were in Edinburgh and Dublin respectively, and the Association visited Cork in 1843. Indeed the choice made by David Brewster (figures 6 and 60) of York for the first meeting was because it was "the most centrical [sic] city for the three kingdoms" [68 p 2, 34], while Morrell and Thackray [67, p 99] describe Oxford, Cambridge, Edinburgh and Dublin as "the four great academic centres of Britain".

Once the point is suggested that Irish and Scottish scientists were particularly prominent for the greater part of the nineteenth century (and we may agree with Herries Davies that there was a decline in the last twenty years), it is not hard to suggest a reason. By the beginning of the nineteenth century, university education had been established for many centuries at four cities in Scotland, and while the history of the institutions had been turbulent, there had been considerable achievement, irregular as it had been. Vivian Green [69] has written of these universities in the eighteenth century that: "While these universities were not free from the abuses seemingly endemic in the ecclesiastical and intellectual life of eighteenth-century Britain, such as nepotism, patronage, facetiousness and neglect of duty, as centres of learning and intellectual activity they could claim to be superior to Oxford and Cambridge."

Figure 60
David Brewster

Figure 61
John Henry Poynting

Figure 62
Andrew Gray, Kelvin's successor in Kelvin's lab c. 1900

Though many of these abuses remained into the nineteenth century, and in Glasgow were strenuously opposed by Kelvin's father and Kelvin himself, there was further progress, and at the very least reasonably stable and secure conditions in which excellent scholarship and research was possible. In Dublin too, TCD was well-established, and, for all Fitzgerald's frustrations with the slow pace of curriculum reform and for the lack of support for applied research, there was a clear opportunity for him and others to carry out productive work.

The situation in England and Wales, considering its far greater population, was far patchier. There were, of course, Oxford and Cambridge, where the level of achievement was high if not necessarily consistent. They will be discussed in the following section. There were developments in London and Durham in the first half of the nineteenth century, and then the steady foundation of the civic or "redbrick" universities in the second half of the century and into the twentieth. Despite the good intentions of the founders, which outside London and Durham, where there were special agendas [69], were usually to provide well-qualified manpower to aid the industrial development of the various regions, progress was slow. Without support from government, there was in most cases little enough money for buildings and a bare complement of staff, and none for personnel or facilities for research. Bristol University, for example, survived for decades on support from the Wills tobacco family.

When Sheffield University applied to join Manchester, Liverpool and Leeds as a member of the Victoria University in 1898, it was turned down on the grounds that "its facilities were inadequate, its premises poor, its finances unbalanced and its pupils of mediocre quality", a judgement which Green [69, p 116] considers unfair, not because it was untrue, but only because "there was hardly a university outside of Oxford and Cambridge of which this was not true at some time in the nineteenth century." It is scarcely surprising that the level of scientific achievement at the civic universities was extremely low until the last decades of the nineteenth century when, with a small amount of support from government, conditions improved marginally and such men as Oliver Lodge (figure 22) and John Poynting (1852-1914) (figure 61) performed excellent work.

Incidentally it is of interest that Kelvin strongly supported independent rather than federal universities. Shortly before his death in 1907, he argued [1, p 1205-6] that QCB should become independent, which did happen the following year, and he added that QCC and QCG should also be given their independence and that the Victoria University should likewise be broken up. As early as 1885, he had supported his former assistant, Andrew Gray, by then Professor of Physics at the University College of North Wales at Bangor, by opening Gray's new physical laboratories and giving an address [1 p 845]. Gray later became Kelvin's successor (figure 62). It is clear that Kelvin had a great faith in university education despite the general financial difficulties of at least the newer universities during his lifetime.

Through the nineteenth century, though, it seems quite natural that Ireland and Scotland could make a substantial contribution to physics and mathematics in the British Isles. The anomaly appears to be the Queen's Colleges, and in particular QCB, which one might expect, at first sight, as a new foundation to be broadly equivalent to the English and Welsh civic universities. Yet, as we have seen, QCB provided a substantial contribution to the development of physics and mathematics, much of it as early as the 1850s. Andrews was there, of course, from the outset, Tait and James Thomson were both appointed in 1854, and Perry was a pupil of Thomson, while Everett came in 1867. Relationships were important: Thomson helped Andrews to analyse his experiments; Tait assisted Andrews in their joint and important experiments on ozone, and warmly acknowledged Andrews' guidance as he moved from mathematics towards physics [11 pp 12-3]; Perry described Thomson as "one of the kindest of men with a strong sense of duty" [40].

The reason for the good showing of the Queen's Colleges and QCB in particular may be that, since they were essentially government foundations, there was far more stability than for the English civics. The official history of QCB [20] does not suggest that finances were strong and professors happy with their salaries, indeed quite the reverse [20 pp 146-7], but one would suspect that the situation was still sufficiently favourable compared to the rather hand-to-mouth existence of the English civics that professors could concentrate on academic studies to a greater extent. And the appointment of Andrews from the start was certainly a master-stroke.

3. Kelvin: Cambridge, Oxford, London and abroad

It has been suggested that, at least for most of Kelvin's life, the greatest opportunities for successful scientific work in Britain were in Ireland and Scotland on the one hand, and Oxbridge and London on the other. In the case of London, we refer not so much to the nascent colleges as to the central institutions, in particular the Royal Society and the Royal Institution. In this section we discuss principally Kelvin's negative, or at least ambivalent, attitude towards Oxbridge and London.

We start with Cambridge and London. Cambridge, of course, helped Kelvin to become a great mathematician. It may be easy to lose sight of this in the light of his later achievements in science and engineering. While there was abundant testimony to his mathematical ability during and soon after his Cambridge years, independent support may be useful in the person of twentieth-century mathematician and historian of mathematics, Clifford Truesdell (figure 63) [70]. Truesdell notoriously insisted that a thermodynamic proposition had not been demonstrated until it had been proved with full rigour; a few central ones, he decreed, had had to wait till the twentieth century and the work of Truesdell himself.

Figure 63
Clifford Truesdell

Figure 64
Joseph Fourier

Figure 65
John William Strutt
3rd Lord Rayleigh

Figure 66
JJ Thomson

Truesdell was contemptuous of the mathematical ability of some of the famous names from the history of thermodynamics. Of Clausius, he (p 206) said that "Few mathematical physicists have shown so little sense of the right mathematics for the job." He remarked that Joseph Fourier (figure 64), by most people considered an immensely important mathematician, and certainly a crucial inspiration for Kelvin, was "boastful and discursive" (p 53); that "Of creative conceptual mathematics there is no example in Fourier's book" (p 77); and that "Among mathematicians commonly reputed great, Fourier sets a record [for *lack of* quality]." Truesdell is by no means uncritical of Kelvin's actual work on thermodynamics. He reports (p 168) that Kelvin's first main paper, published in 1849, was "enthusiastic, rambling, vague". Of his second paper of 1851, Truesdell (p 224) writes that it is "verbose and rambling" and that "this paper, like Kelvin's first one, disappoints a critical student. In its vacillation and obscurity the author of Kelvin's masterful and brilliant researches on electrostatics, hydrodynamics and elasticity can scarcely be recognized (p 234)." Even the ultra-critical Truesdell, however, was forced to admit (p 168) that "Kelvin was an expert and creative mathematician of the highest order".

So Kelvin owed much to Cambridge. He had enjoyed the music and rowing as well as the mathematics. He was a special Life Fellow of Peterhouse from 1872, and was particularly generous towards his old college. There can be no doubt that he enjoyed his visits back to Cambridge particularly perhaps in his earlier years back in Glasgow. Yet it is crystal clear that he regarded the possibility of a permanent return there with disdain. He was offered the Cavendish chair there three times, when it was founded in 1871 and also in 1879 and 1884. On each occasion he rejected the offer, the eventual occupants being Maxwell, Lord Rayleigh (figures 65 and 71) and JJ Thomson (figure 66) respectively.

Figure 67
John Pringle Nichol
[www.universitystory.gla.ac.uk; Glasgow University Archive Services]

Figure 68
Arthur Cayley

One may begin to see his reasons in a letter he wrote to Stokes in 1859 at a time when Kelvin was desperate to get Stokes to come to Glasgow to succeed John Nichol (figure 67) as Professor of Astronomy. As well as occupying his chair at Cambridge, Stokes was also Professor at the School of Mines and was Secretary of the Royal Society. Kelvin wrote of London as well as Cambridge as "those great Juggernauts under which so much potential energy for original investigation is crushed". (Stokes did not apply for the job, because as a devote Anglican, he did not feel free to declare himself a Presbyterian in the still-existent religious test, even though Kelvin assured him the test had become little more than a matter of form.)

Another aspect of Kelvin's ambivalent, even negative, feelings towards working in Cambridge may be seen in his remarks about the work of Arthur Cayley (1821-95) (figure 68) [71]. Cayley had been a contemporary and friend of Kelvin's at Cambridge. Despite being a brilliant mathematician, he was unable at first to obtain a suitable position and worked for 15 years as a lawyer before being invited to become the first occupant of the Sadleirian Chair of Pure Mathematics at Cambridge. His work on groups, matrices and invariants won him everlasting fame, but no plaudits from Kelvin who wrote "Oh that the CAYLEYS would devote what skill they have to such things [elasticity] instead of pieces of algebra which possibly interest four people in the world, certainly not more…..It is really too bad that they don't take their part in the advancement of the world."

Nineteenth-century Cambridge was, of course the home of the mathematical tripos, which was devoted to what today might be called applied and applicable mathematics. Applied mathematics may be defined as the mathematical solution of physical problems, and applicable mathematics as mathematics created specifically for the solution of such problems. Yet, as Warwick [46] says, Kelvin was probably correct to detect and criticise early signs of a growing interest on pure mathematics, study of mathematics for its own intrinsic interest. Indeed such interest was clear in the foundation of the Sadleirian Chair; by the end of the century it would have diluted the thrust to application in the tripos, and in the twentieth century pure mathematics would become probably the most prestigious area of mathematical study and research.

Having studied Kelvin's approach to Cambridge and London, it is natural to move on to Oxford [72] but it is well-known that physics at Oxford was weak through the nineteenth century, and indeed at least to the 1920s or 30s. For the later part of this period it is natural to put much of the blame onto Robert Clifton (1836-1921) (figures 69, 71), Professor of Experimental Philosophy from 1865 to 1919, whose contribution to research in that lengthy period might be described as negligible, perhaps even negative. Clifton was appointed to the position with high hopes, good references and, according to Gooday [73], a "warm rapport" with Kelvin. From the point of view of this paper, it is interesting that one of the main events that helped to destroy these hopes was brought about by two of Kelvin's followers, John Perry (yet again!) and William Ayrton.

In fact Clifton had an early great success in achieving the construction of the Clarendon Laboratory for physics in 1870. This actually predated Cambridge's Cavendish Laboratory by 4 years, and in fact, when helping to design the Cavendish, Maxwell used plans of the Clarendon given to him by Clifton [73]. Indeed, as Graeme Gooday [73] points out, though even at the time it would have been well-understood that Maxwell's achievements in research were vastly greater than those of Clifton, it is not necessarily the case that he was initially more successful than Clifton across the board of the tasks involved in running a new laboratory – effective teaching, encouragement of research, staff training and so on. It would seem that both struggled.

In 1873, however, Clifton was guilty perhaps of rather arrogant behaviour, which must often have come back to haunt him in future years. In June, Perry applied for a fellowship at Clifton's own college, Merton, and was informed that he was indeed eligible. However when he arrived for the examination in October, he found out that his QCB MA was not sufficient to allow him to take the examination, and he wrote a letter [74] published in *Nature* criticising the college for misleading him, and also for refusing him access to the apparatus of the Clarendon to prepare for the examination. Clifton's response [75] was somewhat patronising, and Perry's final response [76] clearly highly frustrated and extremely resentful of Clifton. Certainly Perry would be hoping to get his own back.

The opportunity came sooner than he might have dared to expect. At the time there was much discussion over the source of voltaic action, and in particular a debate between Kelvin and Maxwell. Ayrton and Perry were at the time in Japan, and in 1875-6 carried out an extensive experimental investigation of the topic, the results of which, they claimed, strongly supported the position of Kelvin. They sent a draft to Kelvin himself, who read it at the 1876 BAAS meeting in Glasgow.

Figure 69
Robert Bellamy
Clifton
[Clarendon Laboratory]

Figure 70
Frederick Lindemann,
Lord Cherwell
[Clarendon Laboratory]

Journal of Physics: Conference Series **158** (2009) 012006 doi:10.1088/1742-6596/158/1/012006

Figure 71
Board of Visitors, Royal Observatory Greenwich, 1894
From left, Clifton 1st, Stokes 4th, Rayleigh 5th, Rosse 9th
[Museum of History of Science, Oxford]

Meanwhile Clifton, in gaps between teaching duties, had carried out experiments on the same topic, though unfortunately not at the same level of sophistication as the Japanese ones. Clifton had unresolved problems with aspects of his experiment, but nevertheless in May 1877 sent an interim account to the Royal Society, expressing the hope that, at a later date, he would be able to send a more satisfactory and complete paper. He had been elected FRS in 1868, presumably out of respect for his Oxford chair rather than his very limited research achievements. (Indeed, despite these achievements remaining low, he was later a member of the Council of the Royal Society, and actually Vice-President from 1896 to 1898 [73 p 113].) So it was not surprising, though unfortunate for all concerned (except perhaps Perry), that his paper was speedily published [77]. The paper did not refer to the Japanese work for the very good reason that Clifton had not attended the Glasgow meeting and was unaware of it.

When Ayrton and Perry saw this paper in December of 1877, they apparently thought at first that Clifton was attempting to claim priority over their own paper, but also came to the conclusion that his paper displayed misunderstanding of practical matters and evidence of poor experimental technique. (Clifton, like Kelvin and Tait, had made the move from mathematical studies as a Cambridge wrangler to experimental work. Five years as the first Professor of Natural Philosophy at Owens College, Manchester, had made him a skilled and painstaking user of relatively standard equipment, but probably not as adept at the design and management of challenging experiments.)

Ayrton and Perry quickly sent a paper to the *Telegraphic Journal* [78], which not only emphasised their own priority, but also included a sustained and savage criticism of Clifton's work. They followed this up with an account of their own work [79] that did not deign to cite Clifton's paper at all. As Gooday says [73], it is unsurprising that Clifton never published a research paper again, and thus any hope that the Oxford physics might rival that at Cambridge disappeared for many decades. Much later in 1903, Perry himself was to launch a savage and sustained broadside (around 11 thousand words!) against what he considered to be the highly negative attitude displayed to science at Oxford [80].

It should be mentioned that, as we saw in section 1, Townsend was appointed to a second chair of physics in 1900. Though must of his energies appear to have been dissipated in quarrelling with Clifton, until the First World War his work was not unsuccessful. Indeed Lelong [48] is prepared to talk not only of "Townsend's growing prominence" but also of "the fall of JJ Thomson as the international expert in discharge physics". However after the war Townsend completely failed to keep up with modern physics, relativity and particularly quantum theory, and for the last twenty or so years of his career up to 1941, when he was "asked to leave", his laboratory was perhaps as big an embarrassment to Oxford as Clifton's had been. Clifton's successor from 1919, Frederick Lindemann (figure 70) [81-3] (later Lord Cherwell and effectively Winston Churchill's scientific advisor during the Second World War) was able to persuade a few eminent scientists to work in the Clarendon between the wars, though they remained very much individuals. It was not until after the Second World War that Oxford physics had anything like the strength in depth to be expected in a major university.

Figure 72
Hermann von Helmholtz

Figure 73
Sadi Carnot

Figure 74
Josiah Willard Gibbs

Figure 75
George Green

Having briefly discussed the views of Kelvin and his supporters on the state of physics and mathematics in the various parts of the British Isles, we may even more briefly mention their response to competition or suspected competition from outside. For Europe, we mention the three most important examples [1, 2, 29]. Hermann Helmholtz (1821-94) (figure 72) is today recognised as one of the foremost among the founders of the idea of energy and its conservation. He owed much to early recognition by Kelvin, and the two remained on close terms. One might suspect, though, that this relationship was conditional on Kelvin's part to Helmholtz refraining from over-asserting his own claims. Julius Robert Mayer (1814-78) also made early, though perhaps somewhat incoherent, suggestions in this area. In 1862, Tyndall took the opportunity of attacking Kelvin's group by claiming priority for Mayer over Joule in the matter of the equivalence of heat and work. Tait in particular responded with the greatest vigour [29, Ch 9].

We now turn to Clausius. Today, though it would be recognised that many scientists played a role in the understanding of thermodynamics, and Kelvin and Rankine played crucial roles, it was Clausius who actually achieved the synthesis first. Indeed, in his earliest papers on thermodynamics, Kelvin gave him reasonable credit, dividing that for the Second Law between Sadi Carnot (figure 73) and Clausius himself. In 1868, though, in his book *Thermodynamics* [84], Tait rewrote history by taking Clausius' term "entropy" and applying it to an entirely different quantity, the available energy. In Maxwell's *Theory of Heat* [85] of 1871, the role of Clausius was reduced to the extent that he was forced to complain that "[M]y writings are left quite unmentioned; and my name occurs only once, when it is said that I introduced the word *entropy*; but it is added that the theory of entropy had already been given by W. Thomson. Hence any one who derives his knowledge of the matter solely from this book must conclude that I have contributed nothing to the development of the mechanical theory of heat" [86, 29 p 257]. For the next few years Tait and Clausius continued to bicker over Clausius' contributions to thermodynamics, Tait not only questioning Clausius' priority, but criticising the content and style of his work.

Overall we may surmise that Kelvin and his friends were pleased to maintain good relations with scientists from other nations in Europe provided, but only provided, these scientists were prepared to recognise the general priority of their British colleagues.

Very briefly we look even further afield to the American Josiah Willard Gibbs (1859-1903) (figure 74) is today recognised as having instigated very important developments of thermodynamics. Smith [29] reports that Maxwell became an advocate of Gibbs's work (p. 260), but Kelvin strongly disagreed (p. 303).

4. The Kelvin tradition

We have seen that, while we may not *need* to talk of an Irish or Irish-Scottish tradition, Kelvin did have distinct reservations about science as carried out in Cambridge and London (and his allies showed some contempt for what was carried out at Oxford). It may be suggested that Kelvin established and led a tradition that flourished in particular in the industrial cities of Glasgow and Belfast. Though the social position of Fitzgerald and the architectural splendours and ancient traditions of TCD may seem much closer to Oxbridge than to industry, Weaire [36] suggests that, frustrated by the lack of interest in applied science in Dublin, "He might have been happiest in Belfast among the linen factories and shipyard cranes of the Lagan, looking across to the Clyde," so, to that extent, we may add in some aspects of work in Dublin and create an Irish-Scottish tradition centred mainly, though not necessarily exclusively, on Kelvin himself. The suggestion will have much in common with that of Crosbie Smith [29], but may be somewhat wider in scope than his "energy physics", and will be more open to consideration of the Irish element, as discussed for many of those already mentioned, as well, of course, for the importance of the upbringing of Kelvin and his brother.

Aspects of this tradition may include working on problems of practical importance, being prepared to interact with engineers and industrialists, and actually tackling engineering and design problems. It may also include a willingness to tackle major syntheses, and to discuss the fundamental or theological significance of these syntheses, even perhaps to make use of religious metaphor or mental constructs in the creation of these syntheses. However we will not insist on the view of religion necessarily being positive; it is the interest that is relevant.

We may explore these factors in the context of some of Kelvin's most important work. We will start with his work on what today might be described as potential theory, the relationship between charge density, field and potential, which may be described as the basis of electrostatics, but is also relevant to many other areas of physics, the electrostatic charge being analogous, for example, to the mass in gravitational theory. Kelvin was extremely young at the time, at the beginning of the work an undergraduate and at its conclusion a relatively fresh postgraduate; nevertheless he was coming up with the ideas at the same time as, and in some ways surpassing, such mathematical greats as Gauss and Sturm; as it transpired they were all broadly repeating largely unknown work performed over a decade earlier by George Green (figure 75) [87]. It might be said that Kelvin was working within the Cambridge tripos tradition of applying mathematics to physical problems, but perhaps already extending the tradition beyond the more standard topics of mechanics and planetary motion.

The second area where Kelvin's work was of tremendous significance was electromagnetism. It is, of course, well-known that Maxwell was responsible for the laws of electromagnetism, an achievement which has ensured his fame lasting into the twenty-first century. Almost as well-known is the fact that Kelvin rejected its final form; he opposed in particular the inclusion of the displacement current, but also the duality of vectors – **E** *and* **D** for the electrical vector, **B** *and* **H** for the magnetic. This failing has caused Kelvin's reputation to drop significantly. Yet the irony is that Maxwell's work was built so comprehensively on foundations laid by Kelvin that it would scarcely be unjust for the laws to be called the Kelvin-Maxwell or Thomson-Maxwell laws!

The usual storyline is that Coulomb's law was the centre of the mathematical approach to electromagnetism, that Faraday rescued the subject from its over-mathematical state with his pictorial representations of fields, and that finally Maxwell put Faraday's pictures into a more suitable mathematical form. This is certainly not the whole truth, and the missing parts both relate to Kelvin [87].

First, Kelvin was initially highly unimpressed by Faraday's lack of mathematics, but later came to understand and appreciate his approach. He cleared up some of Faraday's own misconceptions. (Faraday at first thought that Coulomb's law meant that lines of force had to be straight, and that the law was therefore disproved by his own experiments; Kelvin explained that theory and experiment were totally in line.) Kelvin then interpreted Faraday's work for the benefit of Gauss and other prominent mathematical physicists. Faraday and Kelvin interacted fruitfully for the rest of Faraday's career; indeed, in the work on fields, Kelvin's role was so significant that historians of science have debated whether Faraday or Kelvin should be regarded as the founder of field theory [88-9].

Secondly, Kelvin taught Maxwell a vast amount about every aspect of electromagnetism, as freely admitted by Maxwell himself [87, 90-91]. And Kelvin's own analysis, based on the mathematical analogy between heat conduction and electrical attraction, took Maxwell a very long way into his own theory.

Here we are less concerned in deciding why Kelvin failed to take the last step while Maxwell was able to do so, than in recognising that our putative Irish-Scottish grouping, including, of course, both Kelvin and Maxwell, was willing and able to apply advanced mathematical methods to the solution of physical problems, and to tackle major syntheses. In the field of electricity and magnetism, Kelvin also applied theoretical knowledge and experimental ingenuity to the solution of a large number of engineering problems, in particular the Atlantic cable, and put great effort into the establishment of a set of coherent practical units [92].

We now turn to the area where Kelvin is probably best-known – thermodynamics. Here we note that many of the characters involved around Kelvin in this area - Rankine, Tait, James Thomson, Jenkin, and also Maxwell, who, although too young to take part in the fundamental work, made great contributions to the follow-up work of kinetic theory, including, of course, consideration of the so-called Maxwell demon – had strong Irish or Scottish connections, or, interestingly, often both. And certainly the work fulfils our criteria of applying mathematical principles to novel physical topics, and of being willing to tackle major syntheses. Indeed, of the three chief founders of thermodynamics, at least at time of the initial flowering of thermodynamics, Clausius had rather narrow interests restricted largely to heat engines [93], and actually, even among our grouping, the important work of Rankine was mostly restricted to molecular models. In immense contrast, Kelvin's work was concerned with the most general aspects of science including conservation, irreversibility, the new and all-important quantity of energy, and effectively the whole history of the universe.

With such a wide compass, it was inevitable that theological matters would be included. This is in accord with one of our suggestions for Irish-Scottish science, a willingness, even a desire, to tackle such issues where appropriate, and it was very much to the taste, also, of Kelvin and his brother James. Through the 1840s [2], the brothers were perplexed by matching theological and physical problems. On the theological side, on the one hand they were convinced that God was the eternal and only creator, and that it was unthinkable that human beings could themselves create or destroy. On the other, there was a strong biblical tradition of a decaying creation, of change, irreversibility, dissipation; mountains were eroded, and animals and human beings died. In a draft of his 1851 paper, Kelvin gave a reference to the biblical text from Isaiah 51: 6: "For the heavens shall vanish away like smoke, and the earth shall wax old like a garment."

On the physical side, though Newton had believed that there was a requirement for God to maintain the solar system in its many rotations, at least by the beginning of the nineteenth century it would have been practically taken for granted that the motion of the universe was self-sustaining and there would be no 'winding down'. However in 1833 the discovery that Encke's comet was slowing down suggested that there was a resistive medium in the universe, which would eventually bring all motion to a halt, along, of course, with all life.

The brothers were particularly puzzled by the fact that sometimes effort or the burning of fuel produced useful results in the form of work achieved, suggesting some kind of conservation law, but sometimes not. On a canal, the lock-keeper's effort in raising water may have the useful result of raising the level of the boat; sometimes, though, the water may just cascade over the top of the lock, the effort apparently being wasted. Similarly burning of coal or oil will provide a hot reservoir which may be used to obtain useful work using an engine; however the heat from the reservoir may also be transferred straight to a cold reservoir by conduction, with no work produced.

The brothers' puzzles were resolved in the two laws of thermodynamics, produced by Clausius in 1850 and independently by Kelvin in 1851 (Rankine's contributions being highly significant also). The first related directly to conservation of energy. The second said that, though energy was always conserved, it could be rendered unavailable or 'lost in the material world' in the form of low temperature heat which could not be used to do useful work. This second law thus gave irreversibility in nature, dissipation of energy, an arrow of time, and the possibility of an eventual 'heat-death' when all energy will have been downgraded to heat.

It was Kelvin and Tait in *T and T'* [16] who took on the task of announcing and promotion of the energy paradigm, which provided a new approach to the structure of physical law, largely replacing the older Newtonian paradigm with its emphasis on force and acceleration. In their hands the new synthesis, based on the Laws of Thermodynamics, had the result that physics became, almost by definition, the science of energy. (It may be mentioned that Clausius did not take up the idea of energy until the 1860s, though at that time, as already mentioned, he did introduce the very important quantity of entropy, extremely useful for discussion of the irreversibility of the universe, and indeed the basis of an interesting and important re-statement of the Second Law of Thermodynamics [93].) Thermodynamics was, of course, the basis of many inventions, and, of the Irish-Scottish group, James Thomson and Rankine were particularly involved in the design of new engines [29].

We have seen some of the features of Kelvin's important work, and have suggested how these features might be generalised to some of the putative Scottish-Irish group of scientists and engineers. It is interesting to wonder how much further one might take the argument. For example, for all the mutual antipathy between Kelvin and his friends on the one hand and Tyndall on the other, if, as we have said, one of the criteria of membership is applying ideas concerning religion to science, one might say that Tyndall's negative views on religion are as important in his approach to science as the positive views of Kelvin. Thus it may seem possible, reluctantly or otherwise, to include Tyndall.

Stokes, of course, has a foot in both the Irish and the Cambridge camps. He was intensely interested in the relation between religion and science, but it might be said that ideas on religion did not actually play a role in his construction of theories. He seems also to have been also less willing to attempt to construct large-scale syntheses than some members of the group, more willing to stick to the solution of individual problems. McMillan [28] has also pointed out that Stokes did not charge for his advice, for example, to the Grubbs, in rather a contrast to Kelvin, who was convinced of what may be called the Presbyterian doctrine that one should do good work, and where appropriate give no apology for being well paid for it. Indeed, for all their closeness it must be remembered that Kelvin and Stokes were quite different in many ways. One may thus feel inclined to exclude Stokes from our grouping. Consideration of the rest of those mentioned in the first section may be left as an exercise for the reader!

As a last point, we ask the awkward question – should Kelvin actually be regarded as Scottish or Irish? He spent, of course, 75 years in Scotland, but his roots and early upbringing were in Ireland. Indeed in the Home Rule crisis, described by Iain Hutchison in this volume [94], he always claimed he was speaking "as an Irishman", and also spoke of "my beloved Ireland". It is here suggested that by far the most reasonable answer to our question is that he and his family were Ulster-Scots, in the sense that they were equally at home in the North-East of Ireland and the South-West of Scotland. It was natural for his forbears to have travelled from Scotland to make their home near Ballynahinch. Equally it was natural for Kelvin's father to have attended University in Scotland and then to have taken up his position back in Belfast. It was natural for him to have moved to Glasgow after the death of his wife and for academic promotion, but just as natural for his daughter Anna to marry and move back to Belfast with her husband, and to coax her brother James back to Belfast and eventually to his position at QCB. James, of course, was also to move back to Glasgow. Kelvin campaigned for a political position favoured in particular in the North-East of Ireland (although he himself would not have argued for a separation between the various parts of Ireland; he was no advocate of partition), his area of campaigning being the South-West of Scotland.

Particularly after the death of his long-term invalid first wife and when he had purchased the ocean-going yacht, the *Lalla Rookh*, Kelvin frequently used it to travel to Belfast and its surrounding region. He might have planned a scientific discussion, perhaps with Thomas Andrews; maybe some planning of a technological development in Ireland; possibly in the appropriate period a time of political plotting, berating the perceived wickedness of the Home Rulers; and not least a meeting with members of his family.

Glasgow and Scotland doubtless have an excellent claim to regard Kelvin as one of their own; there is no reason why Belfast and Ulster may not do just the same.

Acknowledgement
I would like to thank Joan Whitaker, Subject Specialist in the Science Library at Queen's University Belfast, who has provided much help and guidance in tracing books, papers and pictures. Also I would thank Shan McAnena and Anna Patrick of the Naughton Gallery at Queen's, and Ivan Ewart, Senior Photographer in Media Services at Queen's for their help in providing images.

References

[1] Thompson S P 1910 *The Life of William Thomson: Baron Kelvin of Largs* (London: MacMillan) (2 vols) p 136
[2] Smith C and Wise M N 1989 *Energy and Empire: A Biographical Study of Lord Kelvin* (Cambridge: Cambridge University Press) p 177
[3] Houston K (ed) 2000 *Creators of Mathematics: The Irish Connection* (Dublin: University College Dublin Press)
[4] McCartney M and Whitaker A (ed) 2003 *Physicists of Ireland: Passion and Precision* (Bristol: Institute of Physics)
[5] O'Hara J G, Ref [4] pp 69-76
[6] McHale D, Ref [3] pp 27-32
[7] O'Connor J J and Robertson E F, Ref [4] pp 61-8
[8] Spearman T D, Ref [3] pp 11-20
[9] Hankins TL 1980 *Sir William Rowan Hamilton* (Baltimore: John Hopkins)
[10] Ref [1] p 188
[11] Knott C G 1911 *Life and Scientific Work of Peter Guthrie Tait* (Cambridge: Cambridge University Press)
[12] Burchfield J D 1975 *Lord Kelvin and the Age of the Earth* (London: Macmillan)
[13] England P, Molnar P and Richter F 2007 John Perry's neglected critique of Kelvin's age of the Earth: a missed opportunity in geodynamics *GSA Today* **17** (1) 4-9
[14] Hofmeister A M and Criss R E 2007 Comment [on Ref [13]] *GSA Today* **17** (7) 10
[15] England P, Molnar P and Richter F 2007 Reply [to Ref [14]] *GSA Today* **17** (7) 11
[16] Thomson W and Tait P G 1883 *Treatise on Natural Philosophy* (Cambridge: Cambridge University Press)
[17] Ref [1] pp 452-3; Ref [2] p 364
[18] Burns D T, Ref [4] pp 77-84
[19] Ref [2] pp 111-3
[20] Moody T W and Beckett J C 1959 *Queen's, Belfast: The History of a University* (London: Faber and Faber)
[21] Wood A, Ref [3] pp 33-8
[22] Wood A, Ref [4] pp 85-94
[23] Wilson D B 2003 Arbiters of Victorian science: George Gabriel Stokes and Joshua King *From Newton to Hawking: A History of Cambridge University's Lucasian Professors of*

Mathematics ed K C Knox and R Noakes (Cambridge: Cambridge University Press) pp 295-342
[24] Wilson D B 1987 *Kelvin and Stokes: A Comparative Study of Victorian Physics* (Bristol: Adam Hilger)
[25] Wilson D B (ed) 1990 *The Correspondence between Sir George Gabriel Stokes and Sir William Thompson Baron Kelvin of Largs* (Cambridge: Cambridge University Press) (2 vols)
[26] McMillan N D, Ref [4] pp 95-105
[27] Brock W H, McMillan N and Mollan C (eds) 1981 *John Tyndall: Essays on a Natural Philosopher* (Dublin: Royal Dublin Society)
[28] McMillan N 2000 Ireland and the reform of the politics and government of British science and education, *Prometheus's Fire* ed N McMillan (Carlow: Tyndall Press) pp 481-524. This article contains interesting information about Tyndall's interaction with Stokes.
[29] Smith C 1998 *The Science of Energy: A Cultural History of Energy Physics in Victorian Britain* (London: Athlone) pp 177-91
[30] McMillan N D, Ref [4] pp 106-15
[31] J W J 1905 *Proceedings of the Royal Society* **75** 377-80 [Obituary of Everett]
[32] Ref [24] pp 53-4
[33] Everett J D 1875 *Units and Physical Constants* (London: Macmillan); reissued in 2007 by Kessinger Publishing in their Legacy Reprint Series
[34] O'Connor J J and Robertson E, Ref [4] pp 135-42
[35] Weaire D 2008 Kelvin and Fitzgerald: great Irish physicists *Kelvin: Life, Labours and Legacy* ed Flood R, McCartney M and Whitaker A pp 86-93
[36] Weaire D and Coey J M D 2007 Mentor and constant friend: the life of George Francis Fitzgerald (1851-1901) *European Review* **15** 513-21
[37] Hunt B 1991 *The Maxwellians* (Ithaca: Cornell University Press)
[38] Hunt B J 2007 'Our friend of brilliant ideas': G. F. Fitzgerald and the Maxwellian circle *European Review* **15** 531-44
[39] Fitzgerald G F 1899 Biographical sketch of Lord Kelvin *Lord Kelvin, Professor of Natural Philosophy in the University of Glasgow 1846-99* (Glasgow: James MacLehose and Sons), pp 1-29
[40] Crossland B 2008 *The Lives of Great Engineers of Ulster* vol 2 ed B Crossland and J S Moore (Belfast: NE Consultancy and The Bookshop at Queen's)
[41] H T 1926 *Proc Roy Soc A* **111** iii-vii [Obituary of Perry]
[42] Latimer C 2008 Kelvin and the development of science in Meiji Japan *Kelvin: Life, Labours and Legacy* ed R Flood, M McCartney and A Whitaker pp 212-23
[43] von Jezierski D 2000 John Perry and Alvin Walther, two pioneers of engineering mathematics *Slide Rule Gazette* **1** (Autumn) 60-4, also at http://sliderules.lovett.com/uksrc/Perry%20and%20Walther%20Gaz%202.htm
[44] Flood R, Ref [4] pp 151-60
[45] Warwick A 2003 'That universal aethereal plenum': Joseph Larmor's natural history of physics, *From Newton to Hawking: A History of Cambridge University's Lucasian Professors of Mathematics* ed K C Knox and R Noakes (Cambridge: Cambridge University Press) pp 343-86
[46] Warwick A 2003 *Masters of Theory: Cambridge and the Rise of Mathematical Physics* (Chicago: University of Chicago Press)
[47] McCartney M, Ref [4] pp 168-75
[48] Lelong B 2005 Translating ion physics from Cambridge to Oxford: James Townsend and the Electrical Laboratory 1900-24 *Physics in Oxford 1839-1939* ed R Fox and G Gooday (Oxford: Oxford University Press) chapter 6
[49] Grattan-Guinness I 1997 *History of the Mathematical Sciences* (London: Fontana)
[50] Ref [1] p 277

[51] Whitaker A, Ref [4] pp 52-60
[52] Strunz B, Ref [4] pp 33-43
[53] Gow R, Ref [3] pp 47-53
[54] O'Hara J G, Ref [4] pp 126-34
[55] O'Hara J G, Ref [4] pp 44-51
[56] McMillan N D 1990 Organisation and achievements of Irish astronomy in the nineteenth century – evidence for a 'network' *Irish Astronomical Journal* **19** 101-18
[57] Whyte N 1999 *Science, Colonialism and Ireland* (Cork: Cork University Press) Ch 3; see also Whyte N 1995 'Lords of the ether and light': the Irish astronomical tradition of the nineteenth century *Irish Review* **17/18** 101-18
[58] Ref [29], p 289.
[59] The term 'West British' used by one Irish person about another is always highly pejorative, implying a preference for British rather than Irish culture, and perhaps suggesting sympathy with British political institutions. See, for example James Joyce's story 'The dead' in *Dubliners* (first published in 1914; many subsequent editions). 'North British' in the mouth of a Scot may have similar though less pronounced implications, but appears to be used by English people, particular in commerce, with a neutral meaning. John Wilkes (1725-97), the well-known English political radical, called his paper *The North Briton* to satirise the Scottish connections of the Earl of Bute, the Prime Minister; articles in this paper led to his imprisonment in 1768. See Sainsbury J 2006 *John Wilkes: the Lives of a Libertine* (Aldershot: Ashgate). Smith [29] mentions the *North British Daily Mail* and the *North British Review.*
[60] In a more recent publication, based on a talk on Tait at QCB in a meeting celebrating 'Queen's Thinkers', Smith more explicitly includes Ireland in 'North Britain'. As well as the title of the talk (see below), he writes that 'These networks also show just how closely intertwined were the cultural contexts of Scottish universities with those of Belfast educational establishments' and '[Tait] formed an integral part of a North British scientific intelligentsia, whose members often circulated around academic posts and institutions in Ulster and in Scotland.' Smith C 2008 P. G. Tait, Queen's College and Ulster-Scots natural philosophy *Queen's Thinkers: Essays on the Intellectual Heritage of a University* (Belfast: Blackstaff) ed A Jackson and D N Livingstone pp 7-17
[61] Flood R 2006 Mathematics in Victorian Ireland *BSHM Bulletin: Journal of the British Society for the History of Mathematics* **21** (3) 200-11
[62] McKeith N E, Ref [4] pp 24-32
[63] Weaire D and Coonan S, Ref [4] pp 143-50
[64] Weaire D and O'Connor S 1987 Unfulfilled renown; Thomas Preston (1860-1900) and the anomalous Zeeman effect *Annals of Science* **44** 617-44
[65] O'Hara J G, Ref [4] pp 161-7
[66] Herries Davies G L 1985 Irish thought in science *The Irish Mind: Exploring Intellectual Traditions* (Dublin: Wolfhound) ed R Kearney pp 294-310
[67] Morrell J and Thackray A 1981 *Gentlemen of Science: Early Years of the British Association for the Advancement of Science* (Oxford: Clarendon)
[68] Morrell J and Thackray A (ed) 1984 *Gentlemen of Science: Early Correspondence of the British Association for the Advancement of Science* (London: Royal Historical Society, Camden Fourth Series)
[69] Green V H H 1969 *The Universities* (Harmondsworth: Penguin)
[70] Truesdell C 1980 *The Tragicomical History of Thermodynamics 1822-1854* (New York: Springer-Verlag)
[71] Crilly T 2005 *Arthur Cayley: Mathematician Laureate of the Victorian Age* (Baltimore: John Hopkins)

[72] Fox R and Gooday G (ed) 2005 *Physics in Oxford 1839-1939* (Oxford: Oxford University Press)
[73] Gooday G 2005 Robert Bellamy Clifton and the 'depressing inheritance' of the Clarendon Laboratory 1877-1919 *Physics in Oxford 1839-1939* ed R Fox and G Gooday (Oxford: Oxford University Press) chapter 3
[74] Perry J 1873a Oxford physical science fellowships *Nature* **8** 506
[75] Clifton R B 1873 Oxford science fellowships *Nature* **8** 528
[76] Perry J 1873b The Oxford science fellowships *Nature* **8** 549-50
[77] Clifton R B 1877 on the difference of potential produced by the contact of different substances *Proceedings of the Royal Society of London* **26** 299-314
[78] Ayrton W E and Perry J 1878a The contact theory of voltaic action *The Telegraphic Journal* **6** 99-100
[79] Ayrton W E and Perry J 1878b Contact theory of voltaic action Parts I and II *Proceedings of the Royal Society* **27** 196-238
[80] Perry J 1903-4a Oxford and science *Nature* **69** 207-14. See also Vernon HM 1903-4 *Nature* **69** 269-70 and Perry J 1903-4b *Nature* **69** 270, both under title: Oxford science
[81] Harrod R F 1959 *The Prof: A Personal Memoir of Lord Cherwell* (London: Macmillan)
[82] Birkenhead Second Earl of 1961 *The Prof in Two Worlds: The Official Life of Professor F. A. Lindemann* (London: Collins)
[83] Morrell J 2005 The Lindemann era *Physics in Oxford 1839-1939* , ed R Fox and G Gooday (Oxford: Oxford University Press) chapter 7
[84] Tait PG 1868 *Sketch of Thermodynamics* (Edinburgh: Edmonston and Douglas)
[85] Maxwell JC 1871 *Theory of Heat* (London: Longmans, Green)
[86] Clausius R 1872 A contribution to the history of the mechanical theory of heat *Philosophical Magazine* **43** 106-15
[87] Whitaker A 2008 Kelvin: the legacy *Kelvin: Life, Labours and Legacy* ed R Flood, M McCartney and A Whitaker (Oxford: Oxford University Press) pp 278-306
[88] Doran B G 1975 Origins and consolidation of field theory in nineteenth century Britain: from the mechanical to the electromagnetic point of view *Historical Studies in the Physical Sciences* **6** 132-260
[89] Gooding D 1980 Faraday, Thomson and the concept of the magnetic field *British Journal for the History of Science* **13** 91-120
[90] Sharlin H I 1979 *Lord Kelvin: The Dynamic Victorian* (University Park; Pennsylvania State University Press)
[91] Moyer D F 1978 Continuum mechanics and field theory: Thomson and Maxwell *Studies in the History and Philosophy of Science* **9** 35-70
[92] Tunbridge P 1992 *Lord Kelvin: His Influence on Electrical Measurements and Units* (Stevenage: Peter Peregrinus)
[93] Smith C 1976 Natural philosophy and thermodynamics: William Thomson and 'the dynamical theory of heat' *British Journal for the History of Science* **9** 293-319
[94] Hutchison I 2009 Lord Kelvin and Liberal Unionism, in this volume

AUTHOR INDEX

Institute of Physics
Dirac House, Temple Back
Bristol BS1 6BE UK

CURRAN ASSOCIATES INC
proceedings.com